Electronics Made Simple

Made Simple Books

Accounting
Advertising
Auditing
Book-keeping
Business and the European
   Community
Business Communication
Business Environment, The
Business Law
Business Planning
Business Studies
Economics
English for Business
Financial Management
Human Resource Management
Information Technology
Keyboarding and Document
   Presentation
Law
Management Theory and Practice
Marketing
Office Procedures
Organizations and Management
Philosophy
Psychiatry
Psychology
Sociology
Social Services
Spreadsheet Skills (Excel)
Spreadsheet Skills (Lotus)
Statistics for Business
Teeline Shorthand
Wordprocessing Skills
   (WordPerfect)

Mathematics

Complex Numbers
Differentiation
Integration

Languages

French
Follow-up French (and cassette)
German
Italian
Spanish

Business French (book and
   cassettes)
Business German (book and
   cassettes)
Business Italian (book and
   cassettes)
Business Spanish (book only)

Computer books

Access for Windows
AmiPro for Windows
Excel for Windows
Hard drives
The Internet
Lotus 1-2-3 (DOS)
Lotus 1-2-3 (5.0) for Windows
MS-DOS
MS-Works for Windows
Multimedia
Pageplus
Pageplus for Windows
PowerPoint
Quicken for Windows
Windows 3.1
Windows 95
Word for Windows
WordPerfect (DOS)
WordPerfect for Windows

Electronics Made Simple

Ian Sinclair

MADE SIMPLE
BOOKS

Made Simple
An imprint of Butterworth-Heinemann
Linacre House, Jordan Hill, Oxford OX2 8DP
225 Wildwood Avenue, Woburn, MA 01801-2041
A division of Reed Educational and Professional Publishing Ltd

A member of the Reed Elsevier plc group

OXFORD   AUCKLAND   BOSTON
MELBOURNE   NEW DELHI   JOHANNESBURG

First published 1997
Reprinted 1998, 1999

British Library Cataloguing in Publication Data
Sinclair, Ian R. (Ian Robertson)
  Electronics
  1. Electronics
  I. Title
  621.3'81

ISBN 0 7506 2842 1

Set by Graphicraft Typesetters Ltd, Hong Kong
Printed and bound in Great Britain

FOR EVERY TITLE THAT WE PUBLISH, BUTTERWORTH-HEINEMANN
WILL PAY FOR BTCV TO PLANT AND CARE FOR A TREE.

Contents

Preface

The aim of this book is to present an outline of modern electronics with a strong emphasis on understanding how systems *work* rather than on the details of circuit diagrams and calculations. This makes the book ideal for the beginner who all too often finds that the mass of detailed information in textbooks is overwhelming and who wants to find how electronic devices work but who does not yet need to know how to design them or how to construct them. If you need to gain experience in practical electronics, take a look at the book by Keith Brindley called *Starting Electronics*. Appendix A contains a list of books that are useful for any reader who wishes to take the subject further.

The emphasis of this book is therefore on the overall view of electronics, the house shape rather than the bricks and mortar, concentrating on waveforms and graphs along with block diagrams that show the effect of a circuit on its input rather than exploring the way that circuit carries out actions. A few items of fundamental background knowledge, such as the behaviour of electrons, are essential and are described briefly from scratch, not relying on any previous learning from GCSE science modules (since these often have a very small physics content). The mathematical content is minimal, because this book concentrates on behaviour rather than on detailed calculations.

The universal use of integrated circuits (ICs) now makes it pointless to describe detailed circuit action, since only the IC designer knows what goes on inside a chip. The emphasis is now on inputs and outputs and the effect of each part of an electronic device on these signals.

This book is particularly useful for anyone who is, initially at least, more likely to use electronic equipment than to design it. This includes anyone who wants to learn about electronics with a view to a career in electronics, and also to students of other subjects who are likely to be using electronic instruments and methods. The examples of uses of electronics have been confined to consumer devices rather than industrial electronics, because these devices are much more likely to be familiar to readers, so avoiding the need to come to terms with two sets of unfamiliar concepts together. In addition, some indication of how methods have developed over the years has been included to make it easier to understand how we have arrived at the system we use today.

Ian Sinclair

Chapter 1   Waves and pulses

Fundamentals

Electronics is a branch of electrical engineering which is concerned with controlling electrons and other charged particles.

That definition doesn't tell you much unless you know something about electrons and electric charge. No one knows what charge *is*, but we do know a lot about what charge *does*, and what we know about what charge does has been accumulated since the time of the ancient Greeks. We can summarise what charge does (the *properties* of charge):

When you rub two objects together they will both usually become charged. These charges are of two opposite types, one called positive, the other called negative.
  Two charges of the same type (two positives or two negatives) repel each other; two opposite charges (one positive, one negative) attract each other.
  The natural state of any substance is not to have any detectable charge, because it contains equal quantities of positive and negative charges.

What we know about the way that charge behaves has led to finding out more about what it is, and we know now that charge is one of the most important effects in the universe. What we call charge is the effect of splitting atoms, separating small particles called electrons from the rest of the atom. Each electron is negatively charged, and the amount of charge is the same for each electron. The other main part of an atom, the nucleus, carries exactly as much positive charge as the electrons around it carry negative charge, Figure 1.1, if we picture the atom as looking like the sun and its planets. For example, if there are six electrons then the nucleus must carry six units of positive charge.

Modern physics has long abandoned pictures of atoms as sun and planet systems, but this type of picture of the unimaginable is good enough for all our purposes.

When an electron has become separated from the atom it belongs to, the attraction between the electron and its atom is, for such tiny particles, enormous, and all the effects that we lump together as electricity, ranging from lightning to batteries, are caused by these force effects of charge. The forces

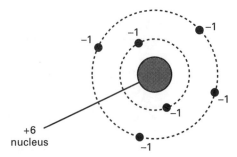

Figure 1.1 A sun and planet view of an atom. Though as a concept this is out of date, it helps to illustrate the idea of electrons whose total charge balances the charge of the nucleus and also the idea that the outermost electrons can be detached.

between charges that are at rest are responsible for the effects that used to be called static electricity (or *electrostatics*), and these effects are important because they are used in several types of electronics devices.

The forces are so enormous that we can usually separate only one electron from a nucleus, and only at enormous temperatures, such as we find within the sun, can all the electrons be separated.

Another option for an electron that has become separated from an atom is to find another atom that has lost its electron (and is therefore positively charged). The movement of electrons from one atom to another causes a large number of measurable effects such as electric current, magnetism and chemical actions like electroplating. Of these, the most important for electronics purposes are electric current, and its effect, magnetism.

The movement of electrons that we call electric current takes place in a *circuit*, a closed path for electrons that has been created using conducting material. All circuits for current are closed circuits, meaning that electrons will move from a generator through the circuit and back to the generator again. This is essential because unless electrons moved in a closed path like this, many atoms would be left without electrons, and that's not a condition that could exist for long.

Electric current is the amount of charge per second that passes a point in a circuit. Electric voltage is the amount of work that each charge can do when it moves.

These are formal definitions. We can't count the number of electrons that carry charge along a wire, and we can't measure how much work is done when a charge moves. We can, however, measure these quantities because of the effects that they have. Current along a wire, for example, will cause a force on a magnet, and we can measure that force. The voltage caused by some separated charges can be measured by the amount of current that will flow when the charges are allowed to move. The unit of current is called an ampere or amp, and the unit of voltage is a volt. The names come from the pioneers Ampère and Volta.

We can create less formal definitions for ourselves. Voltage is like a propelling force for current, and current itself can be thought of as like the current of a river. If we continue with this idea, voltage corresponds to the height of the spring where the river starts.

All substances contain electrons, which are the outer layer of each atom. Some materials are made out of atoms that hold their electrons tightly, and when electrons are moved out of place it is not easy for them to return to their positions. In addition, other electrons cannot move from their own atoms to take up empty places on other atoms. We call these materials *insulators*, and they are used to prevent electric current from flowing. In addition, insulators can be charged and will remain charged for some time. A good example is the party balloon which is charged by rubbing it against a woollen sweater and which will cling to the wall or the ceiling until its charge is neutralised. Surprisingly high voltages can be generated this way on insulators, typically several kilovolts (kV), where kilo means one thousand. For example, 5 kV means five thousand volts. A very small current can discharge such materials, and we use the units microamp (μA), meaning a millionth of an amp, and nanoamp (nA) meaning a thousandth of a millionth of an amp.

Other atoms can be so tightly packed together that they can share electrons, and the loss of an electron does not cause such a large upset in any one atom. In these materials, electric current can flow by shifting electrons from one set of atoms to the next, and we call these materials *conductors*. All metals contain close-packed atoms, and so all metals are conductors. A small voltage, perhaps 1.5 V, 6 V, 12 V or so, can push electrons through a piece of metal, and large currents can flow. These currents might be of several amps, or perhaps smaller amounts measured in milliamps (mA), with the milliamp equal to one thousandth of an amp.

As well as being closely packed, the atoms of a metal are usually arranged in a pattern, a *crystal*. These patterns often contain gaps in the electron arrangement, called *holes*, and holes can also move from one part of the crystal to another (though they cannot exist beyond the crystal). Because a hole behaves like a positive charge, movement of holes also amounts to electric current and in most metals, when electric current flows, part of the current is due to hole movement and the rest to electron movement.

There are some materials for which the relative amount of electrons and holes can be adjusted. The materials we call *semiconductors* (typically silicon) can have tiny amounts of other materials added during the refining process so that we can create a material that conducts mainly by hole movement (a p-type semiconductor) or mainly by electron movement (an n-type semiconductor). In addition, the number of free particles is less than in a metal, so that the movement of the particles is faster (for the same amount of current) than it is in metals, and the movement can be affected by the presence of charges (which attract or repel the electrons or holes and so interfere with movement). The movement can also be affected by magnets. Semiconductors are the way by which electron and hole movement can be controlled.

All materials can be charged, for example, by dislodging electrons. For every positively charged material there is a negatively charged material, because charging is caused by separating electrons from atoms, and the electrons will eventually return. The movement of electrons is what we detect as electric current, and the 'pushing force' for the current, caused by the attraction between positive and negative, is called voltage. Voltage is the cause of current. Materials can be roughly classed as conductors or insulators. Conductors have close-packed atoms which can share electrons, so that electron movement is easy, and electric current can flow even with only a small voltage. Insulators have more separation between atoms, electrons are not shared, and even a very high voltage will not cause any detectable current to flow. When we work with conductors we use low voltage levels and comparatively high currents. When we work with insulators we can use high voltage levels and very low currents.

The third class of material is the semiconductor. Semiconductors can be natural, but better results are obtained by refining materials and deliberately adding impurities that will alter the number of free electrons or holes. The importance of semiconductors is that they make it possible to control the flow of charged particles, and this is the whole basis of electronics.

Before semiconductors were discovered, electronic valves were used to control electron (not hole) flow. The principle, dating from about 1904, is that when electrons move in a vacuum their movement through a wire gauze (or grid) can be controlled by altering the voltage on the grid. Electronic valves are still used where large voltages and currents have to be controlled, such as for radio transmitters, and also for cathode ray tubes (for TV receivers and for the measuring instruments called oscilloscopes), but their use for other purposes has died out.

Steady voltage

There are several ways of generating a steady voltage, but only two are of importance for everyday purposes. Batteries are the most familiar method, and the invention of the first battery by Alessandro Volta in 1799 made it possible to study comparatively large electric currents at voltage levels from 3 V to several hundred volts. A battery is, strictly speaking, a stack of cells, each cell converting chemical action into electrical voltage. In the course of this, a metal is dissolved into a metal salt, releasing energy in the form of electrical voltage that can make current flow, Figure 1.2.

If the energy were not converted into electrical form it would be converted to heat, which is what most chemical changes provide.

Some types of cell are rechargeable, so that you can connect them to a (higher) voltage and convert the metal salt back to the metal – but you need to use more energy than you received out of the cell. As always, no energy is created. If anyone ever tries to sell you a motor that runs on air, magnetism or moonbeams, always ask where the energy comes from (usually it's from the people who have been cheated along the way).

The other way of generating a steady voltage was discovered by Michael Faraday in 1817. He demonstrated the first dynamo, which worked by rotating a metal disc between the poles of a magnet, using the energy of whatever was turning the disc (Faraday's hand first of all, and later a steam engine) into electrical voltage that could provide current, Figure 1.3. Strictly speaking, a voltage that is the result of a generator or a battery should be called an electromotive force (EMF), but this name is slightly old-fashioned nowadays.

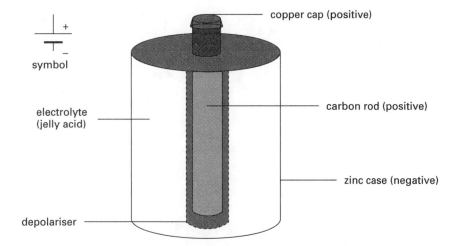

The acid dissolves the zinc case, releasing electrons so that the case is negative. Positive charge accumulates on the carbon rod. The depolariser removes hydrogen gas which otherwise acts as an insulator.

Figure 1.2 A typical cell, the familiar zinc–carbon type. The voltage output is 1.5 V for a fresh cell, and the current that can be drawn depends on the size of the cell, up to a few amperes. The electrical energy is obtained at the expense of chemical energy.

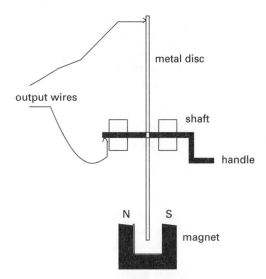

Figure 1.3 Faraday's first electrical generator. When the disc is spun, a voltage appears between the contact points or brushes. The voltage is very small, but the principle can be applied to make a dynamo.

Later, Faraday found that higher voltages could be generated by substituting a coil of wire for a metal disc, and this is the basis for modern dynamos.

In these pioneering days, steady (or direct) voltage was the only type that was thought to be useful. A voltage will cause a current to flow when there is a path of conducting material between the points where the voltage (the EMF) exists – we call these points *terminals*. A cell or a simple dynamo will have two terminals, one positive and the other negative. When a wire or any other conductor is used to connect the terminals, a current will flow, and if the voltage is steady, then the current also will be steady. That does not mean that it will be steady for ever. A cell will be exhausted when its metal is all converted, and the voltage will fall to zero. A dynamo will generate a voltage only for as long as the shaft is turned. By convention, we say that the current flows from the positive terminal to the negative terminal.

A steady voltage will cause a steady current, direct current (DC) to flow, and in 1826 Georg Simon Ohm found what determined how much current would flow. He called this quantity resistance. The three quantities, voltage, current and resistance are therefore related, and the unit of resistance is called the ohm in his honour.

When a current of I amps flows, using a voltage of V volts, then the resistance R ohms is equal to voltage divided by current. In symbols, this is $R = V/I$, and we can write this also as $V = RI$ or $I = V/R$.

This relationship is often, wrongly, called Ohm's law. Ohm's law states that this quantity called resistance is constant for a metal that is used to conduct current at a steady temperature. What this boils down to is that we can use the relationship in any of its three forms with R constant if the resistance R is of a metal at constant temperature. In other words, if 6 V causes 2 A to flow through a resistance, then 12 V will cause 4 A to flow. This is not true when some non-metal materials are used, or if a metal is allowed to get hot. For example, the current through a semiconductor obeys $I = V/R$, but the value of R is not constant, it changes for each value of current. Similarly, a torch-bulb has a higher resistance when it is hot than when it is cold.

We can determine whether a conducting material obeys Ohm's law by plotting a graph of current against voltage, using a circuit like Figure 1.4(a). If this results in a graph that is a straight line (Figure 1.4b), then the material obeys Ohm's law; it is ohmic. Most metals behave like this if their temperature is kept constant. If the graph is curved (Figure 1.4c) the material is not ohmic, and this type of behaviour is found when metals change temperature and when we use semiconducting materials in the circuit.

Changing voltage

The (EMF) voltage that is generated by a cell or battery is truly steady, but the voltage from a dynamo is not. When one side of the revolving coil of wire approaches a pole of the magnet, the voltage is in one direction, but as the wire moves away from the magnetic pole, the voltage is in the other direction. Because there are two poles to a magnet, the voltage from a rotating coil rises and falls twice in a revolution, positive for half the time and negative for the other half. The graph looks like that of Figure 1.5, and the

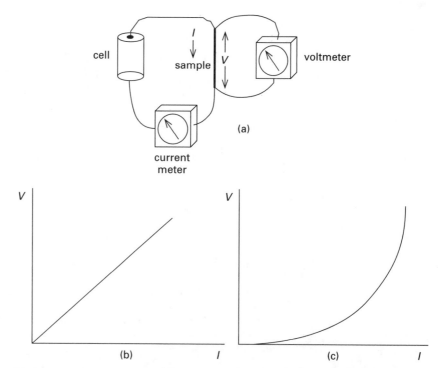

Figure 1.4 Voltage, current and resistance. (a) The voltage (EMF) of a cell is used to pass current through a sample of conducting material. The ratio of voltage across the sample to current through the sample defines the resistance. (b) Ohmic graph, (c) one form of non-ohmic graph.

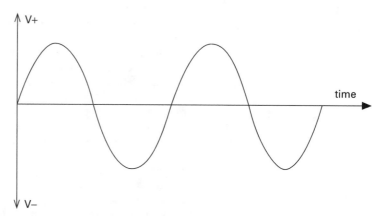

Figure 1.5 A sinewave which can be produced by rotating a coil between the poles of a magnet. This is the form of AC wave used for electrical supplies and also for radio carrier waves (see later). Radio carriers are generated by circuits called *oscillators* rather than by rotating machines.

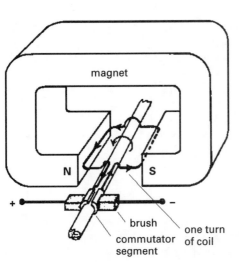

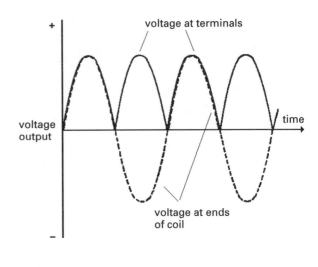

Figure 1.6 Principle of the dynamo. The coil generates a voltage that is alternating, but by reversing the connections on each half turn, the output voltage is in one direction (unidirectional). Modern dynamos use slip-rings and semiconductor circuits (see Figure 2.14) rather than the commutator.

time it takes to go through one complete cycle of the shape is the time that it takes to turn the coil through a complete revolution.

Faraday's later type of dynamo got around this by using a mechanical switch called a *commutator*, which reversed the connections to the coil twice on each revolution, just as the voltage passed through zero, Figure 1.6. This generated an EMF which, though not exactly steady, was at least always in the same direction, so that it was possible to label one terminal as plus and the other as minus. If the shaft of the dynamo is spun fast enough, there is in practice very little difference between the voltage from the dynamo and the same amount of voltage from a battery. For tasks like heating, lighting, electroplating and so on, the supplies are equivalent.

There are other ways of generating electricity from heat and from light, but they suffer the problems of poor conversion (not much electrical energy out for a lot of heat energy in) or low rate of energy (for example, you have to cover a lot of ground with light-cells to generate electricity for a house). A few small generators use nuclear power directly, by collecting the electrons that radioactive materials give out, but large-scale nuclear reactors use the heat of the reaction to generate steam and supply it to turbines. These are therefore steam-powered generators, and the only difference is in how the steam is obtained. The same is true of the places on earth where steam can be obtained from holes in the ground, providing geothermal power.

Voltage (EMF) can be generated by converting energy (mechanical or chemical (into electrical form. A cell converts chemical energy into a steady voltage, but only for as long as there is metal to supply the energy. A dynamo uses mechanical energy and will generate a voltage for as long as the shaft can be turned by a steam engine, wind-power, a water-wheel, or whatever source of energy is at hand. The output of a dynamo is not steady, but is in one direction and can be used for the same purposes as truly steady voltage.

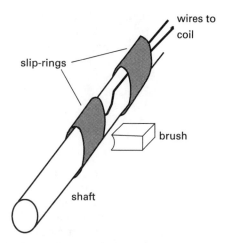

Figure 1.7 Slip-rings provide a way of connecting to each end of a rotating coil.

Waves

Alternating voltages

If we omit the commutator of a dynamo and connect the ends of the rotating coil, the graph of the voltage, plotted against time, shows a shape that looks like a wave, as was illustrated in Figure 1.5. The connection to the coil is made using slip-rings, Figure 1.7, so as to avoid twisting the wires. This voltage is neither steady nor in the same direction all the time, and a voltage like this is called an alternating voltage. When we use an alternating voltage in a circuit, the current is also alternating current, AC. A graph of current plotted against time is of the same shape as the graph of voltage, and the peak of current is at exactly the same time as the peak of voltage in each direction.

In practice, the rotating portion is an electromagnet and the slip-rings carry the steady current that magnetises this rotor. The output is from a coil wound on the non-rotating portion (stator), so that the much larger output current does not have to be passed through a brush and slip-ring. This device is called an *alternator*, and its most familiar form is the generator in a car.

We can use AC for electrical heating, electric light, motors and in fact for most of our domestic uses of electricity. It cannot be used for electroplating, however, and it cannot be used for electronic equipment. In the nineteenth century, however, electronics was in its infancy, and the advantages of AC greatly outweighed any minor disadvantages, particularly since AC could be converted to DC if that was essential. What were the advantages of using AC?

1 AC is the natural output from a rotating generator, requiring no commutators or other reversing devices.
2 Alternating voltage can be converted up or down (using a transformer, see later) without using any mechanical actions. For example, if you generate at 5 kV you can convert this up to 100 kV or down to 240 V with practically no losses. The higher the voltage you convert to, the longer the distance you can connect by a cable of reasonable size. This makes the

National Grid possible, so that generating stations need not be close to users of electricity.

3 Very simple motors can be made that use AC and which will run at a constant speed (used for clocks, gramophone motors and tape recorder motors).

4 AC can be used to power vibrating motors, such as used for electric shavers.

5 AC can be converted to DC for electronic equipment, and can be used to provide several different steady voltage levels for one AC supply.

Virtually every country in the world therefore generates and distributes electricity as AC, and the convention is to use a rate of 50 cycles per second (50 hertz or Hz) or 60 cycles per second. In terms of a simple generator, this corresponds to spinning the shaft of the generator at 3000 r.p.m. for 50 Hz, or 3600 r.p.m. for 60 Hz. The abbreviation Hz is for hertz, the unit of one cycle of alternation per second. This was named after Heinrich Hertz, who discovered radio waves in 1884.

DC is the natural output of a battery, but AC is the natural output of a rotating generator. AC is used world-wide for generating and distributing electricity, mainly because it makes it possible to have a large distance between the generator and the user. Since AC can be converted to DC much more easily than converting DC to AC, there are no problems in using an AC supply for electronics circuits that require a steady voltage supply.

Electromagnetic waves

AC would be important enough if it only provided a way to generate and distribute electrical power, but it has even greater importance. A hint of this came in 1873 when James Clerk Maxwell published a book containing equations that showed that an alternating voltage could generate waves of voltage and magnetism in space, and that these waves would travel at the same speed as light. He called these waves *electromagnetic waves*, and from there it was a short step to show that light was just one of these waves.

Why one? These waves differ from each other in two ways. One is the number of waves per second or *frequency* of the waves. The other is the *amplitude* (the amount of rise and fall), Figure 1.8. Amplitude measures the energy of the wave, so that a large amplitude of light wave means a bright light. Frequency affects how easily a wave is launched into space and how we detect it.

Low frequency electromagnetic waves are called radio waves, and we generate and detect them nowadays using electronic methods. To put figures to these quantities, waves with frequencies below 100 kHz (one hundred thousand complete cycles per second) are classed as very low frequency (VLF) and are used mainly for time signals and for some long-distance communications. Waves of around 1 MHz (one million hertz) frequency are called medium wave, and a large number of entertainment radio transmitters use this range. Waves in the range 10 MHz to around 50 MHz are classed as short waves, used for communications, and the VHF range 50 MHz to 200 MHz is also used for similar purposes. As the frequency is increased, the range of useful communications decreases.

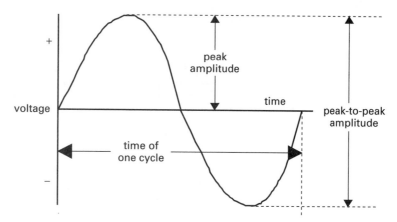

Figure 1.8 Amplitude. The peak amplitude figure is used for symmetrical waves, like sinewaves. Peak-to-peak readings are used for waves with a shape that is not symmetrical (see Figure 1.9b).

The amplitude that is quoted for a wave is usually the peak amplitude. The figure of peak-to-peak amplitude is used when the wave is not symmetrical.

The range of 300 MHz to 1000 MHz is ultra-high frequency (UHF), used for television transmissions, and once we get to using the unit of GHz (1 gigahertz is equal to 1000 MHz) then the signals are in the microwave ranges. These names are only rough indications of a range, and because we have found it necessary to use higher and higher frequencies over the years we have had to invent names for new ranges of frequencies that we once thought unusable.

One range is particularly significant to us and is called the audio range. This is the range of frequencies between about 30 Hz and 20 kHz, and its significance is that this is the range of frequencies of sound that we can hear (bats might have a different definition). A microphone, for example, used in a concert hall would provide an electrical output that would consist of waves in this range. For speech, we can use a much smaller range, about 100 Hz to 400 Hz.

A microphone is an example of an important device called a transducer. A transducer converts one form of energy to another, and for electronics purposes, the important transducers are those which have an input or an output which is electrical, particularly if that input or output is in the form of a wave.

All the electronic methods that we know for generating waves are subject to some limitation at the highest frequencies, and other ways of generating and detecting waves have to be used. At around 1000 GHz, the waves are called *infrared*, and their effect on us is a heating effect; we can generate these waves from any warm object. Higher frequencies, to about 100 000 GHz, correspond to the infrared radiation from red-hot objects, and one small range of frequencies between 100 000 and 1 000 000 GHz is what we call light. Higher still we have X-rays, gamma rays, and others which we find very difficult to detect and cannot generate for ourselves.

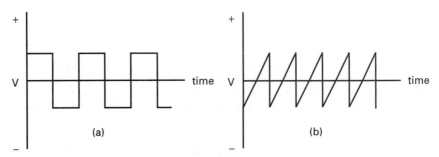

Figure 1.9 Other waveforms (a) square waves, (b) sawtooth or sweep waveform.

As far as electronics is concerned, we make most use of the waves in the range from a few Hz to a few GHz, and radio technology is concerned with how these waves are generated, used to carry information, launched and detected. Radio, in this respect, includes television and cellular telephones, because the use of the waves is the same; only the information is different.

Waveforms

The waveform of a wave is its shape, and because an electromagnetic wave is invisible, the shape we refer to is the shape of the graph of (usually) voltage plotted against time. The instrument called the oscilloscope (see Chapter 13) will display waveforms; we do not need to draw graphs from voltage and time readings. We are seldom particularly interested in the shape of the waves that are transmitted through space, and waveforms are of interest mainly for waves that are transmitted along wires and other conductors in electronic circuits.

The simplest type of waveform is the shape of the wave that is generated by the magnet and coil arrangement that Faraday used (which was illustrated in Figure 1.5). The shape that this generates is called a *sine wave* because it is the same shape as a graph of the sine of an angle plotted against the angle. What makes the sine wave particularly important is that any other shape of wave can be created by mixing sine waves of different frequencies, and any shape of wave can be analysed in terms of a mixture of sine waves.

Though a pure sine wave is the simplest wave, its uses are confined to AC power generation and to radio transmitters. Most of the waves that we use in electronics are a long way from a sine wave shape, and one special type, the pulse, has become particularly important in the second half of the twentieth century.

Take a look at the two shapes in Figure 1.9. Wave (a) is called a square wave, for obvious reasons, and it is used particularly when a wave is used for precise timing. Because the edges of the wave are sharp, each can be used for starting or stopping an action. If you used a sine wave, there would be some uncertainty about where to start or stop, but the steep edge of the square wave makes the timing action much more certain. For example, the time needed to change the voltage level of such a wave might be only 50 nanoseconds (1 nanosecond is a thousandth of a millionth of a second). The square wave is nothing like a sine wave, and it can be generated naturally by switching a steady voltage on and off rather than by any type of rotating generator. The most precise square waves are generated by electronic circuits that use a vibrating quartz crystal to control the frequency, and these circuits are used in clocks and watches.

The other wave, in Figure 1.9b, is called a *sawtooth* in the UK or a sweep

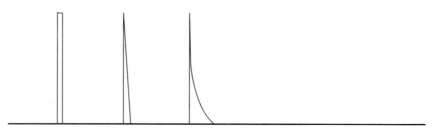

Figure 1.10 Examples of pulse waveforms. The common factor is a sharp leading edge.

in the USA. It features a long even rise (or fall) followed by a fast return to the starting voltage. Before 1936, a wave of this shape would have been an academic curiosity, but this is the shape of wave that is needed for a cathode ray tube, for television or for radar, and we'll look at it again in Chapter 9. The sawtooth is also an important waveform that is used in electronic measuring instruments.

What marks these waves out as totally different from the sine wave is that they show very sharp changes. A sine wave never changes rapidly, its rate of change is fixed by its frequency. These square and sawtooth waves can change in a time that bears no relation to the frequency. A square wave might have a frequency of only 1 Hz, but change voltage in less than one millionth of a second (a microsecond, written as 1 µs).

The waveform is the shape of an electrical wave, and the most fundamental waveform is the sine wave that is generated by a coil rotating between the poles of a magnet. The important features of a waveform are its frequency (the number of times a wave repeats per second) and its amplitude (the height of the wave). Sine waves are used mainly in radio applications, but other wave shapes such as square and sawtooth waves are used, particularly where timing is important.

Pulses

The pulse is a waveform, but one that you might not recognise as a wave because the time of the pulse is very short compared to the time between pulses. Figure 1.10 illustrates three different pulse shapes, all of which share a very sharply-rising portion, the *leading edge*. For a negative pulse, this leading edge would be a sharp fall of voltage.

A pulse is a rapid change of voltage which is of very short duration compared with the time between pulses.

For example, a pulse might repeat at a rate of 1 kHz, 1000 pulses per second. The actual pulse might have a duration, a pulse time, of only 10 µs, so that the change of voltage lasts only for 10 µs in the 1 ms (millisecond) between pulses. That makes the time of the pulse 1/100 of the time between pulses, and these are fairly typical figures. Pulses are used for timing, and they have the advantage that they use very little energy because the change of voltage is so short. A pulse can be used to start an action, to stop an action or to maintain an action (such as keeping a wave in step with the pulses).

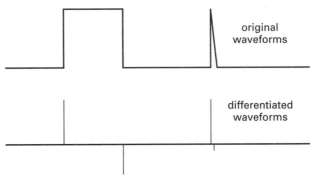

Figure 1.11 Differentiating action, illustrated on two waveforms. The action emphasises the sharply-changing portions of the waves.

Modern digital electronics systems rely heavily on the use of pulses, and when we work on these systems we are not greatly concerned about waveshapes, only about pulse *timing*. There are many things that can change a waveshape, making it very difficult to preserve the shape of a wave. By contrast it is very difficult to upset pulse timing, so that circuits which depend on pulse timing are more reliable in this respect than circuits that depend on waveshapes. The classic example is sound recording. Hi-fi systems in the past tried to work with a waveform that had a very small amplitude (the output from a gramophone pick-up) and keep the shape of the waveform in the form of a copy that had a much larger amplitude. This large-scale copy was used to operate loudspeakers, and we called the whole exercise *amplification*.

The problem with this system is that a copy of the waveform is never perfect, something that is obvious when you make a copy of a copy. Any blemish on the surface of the record, any false movement of the stylus, any interfering signals in the circuits all will make the copied waveform inaccurate, a process we call *distortion*.

Nowadays the sound is recorded as a set of pulses, using the pulses to represent numbers, and each point in a waveform is represented by a number. Using pulses for counting, we can ensure that the numbers are not changed, so that when the stream of numbers is converted back to a wave, the shape of the wave is exactly the same as was recorded. This is the basis of *compact discs*, the most familiar of the digital systems that have over the last few years replaced so many electronics methods. There will be more of all that in Chapters 11 and 13.

Actions on pulses

There are two actions that can be carried out on pulses and on square waves that are important for many purposes. One of these is *differentiation*, and this can be achieved by passing a pulse or square wave into a circuit that selectively passes only the fast-changing part of the input. Figure 1.11 shows the result of differentiating, which converts the pulse or square wave into a pair of sharp spike shapes. These spikes are very short pulses, and we can use them for timing. We can select either a positive or a negative spike by using other circuits.

The other action is called *integration*, and it is a form of averaging or smoothing. A typical action is illustrated in Figure 1.12, showing a set of pulses as the input to an integrating circuit (an integrator). The output is a steady rise in voltage, and eventually this will become steady at a value

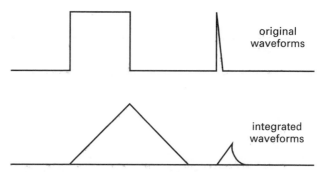

Figure 1.12 Integrating action. The action smoothes out sharp changes, altering a steep rise into a sloping rise, for example.

which is the peak voltage value of the pulses. This is the opposite action to differentiation, removing rapid changes from a waveform.

There is another waveform which is very important in almost all branches of electronics, the *sawtooth* or *sweep* wave. This is obtained by integrating part of a square wave, and was illustrated earlier in Figure 1.9b. The steady rise (or fall) of voltage is called the *sweep* portion, and the rapid return (the portion that is not integrated) is called the *flyback*. We shall meet this type of wave again in connection with television, oscilloscopes and digital voltmeters.

Summary

Pulse and square waveforms can be differentiated or integrated by using suitable circuits. Differentiation emphasises the sharp changes in a wave; integration smoothes out such changes.

Chapter 2   Passive components

Components

Electronic *components* are the building blocks of an electronic circuit, and any electronic circuit is created by connecting components together. At one time this was done by connecting wires between the terminals of components, but nowadays the connections are more likely to be made using metal tracks on an insulating board (a *printed circuit board* or PCB). In such a circuit, both steady and alternating voltages will exist together, and several types of components do not behave in the same way to alternating voltages as they do to steady voltages. In addition, components can be *active* or *passive*. Active components are used to copy (*amplify*) waveforms and to switch voltages and currents on and off under electrical control. Such active components need an *input* signal (a waveform) to control an *output* signal, and they also need some source of power, which is usually a steady voltage supply. A circuit which contains active components can produce an output waveform which provides more power (voltage multiplied by current) than its input waveform. In other words, active components can provide amplification.

Passive components always *reduce* the power of an input waveform, so that an output wave from a circuit that contains only passive components is always at a lower power than the input. Passive components do not need any additional steady voltage supply to enable them to deal with waveforms. A complete electronic circuit will normally consist of both active and passive components, arranged so that the passive components control the action of the active components and act as a path for signal waveforms.

Resistors

Resistors are the most common of passive components. We saw in Chapter 1 that the quantity called resistance is the connecting link between current and voltage, so that in any circuit, the ratio V/I is the resistance. If we measure voltage in volts (V) and current in amps (A), then the resistance is in units of ohms (Ω). Using the Greek letter omega for ohms avoids the confusion that would be inevitable if we used a capital letter O.

A resistor is used to control current or to convert a wave of current into a wave of voltage, using the $I = V/R$ or $V = RI$ relationship.

Every part of a circuit has resistance to the flow of current, and when we specify a resistor as a circuit component we mean a component that is manufactured to some precise (or reasonably precise) value of resistance and used to control the amount of current flowing in some part of a circuit. Though it is possible to manufacture resistors with low values of a few ohms, most of the resistors that we use in electronics circuits have higher values of resistance, and to avoid having to write values like 15 000 Ω or

Table 2.1 Preferred values for 5%, 10% and 20% tolerance

5%	10%	20%		5%	10%	20%
1.0	1.0	1.0		3.3	3.3	3.3
1.1				3.6		
1.2	1.2			3.9	3.9	
1.3				4.3		
1.5	1.5	1.5		4.7	4.7	4.7
1.6				5.1		
1.8	1.8			5.6	5.6	
2.0				6.2		
2.2	2.2	2.2		6.8	6.8	6.8
2.4				7.5		
2.7	2.7			8.2	8.2	
3.0				9.1		

2 200 000 Ω we use the letter k to mean 'thousand' and M to mean 'million' and omit the omega sign. In addition, the letter R is often used to mean ohms, because typewriters (which, unlike word-processors, do not have the omega symbol) are still being used. Another way of making values clearer is to use the letter R in place of a decimal point, because decimal points often disappear when a page is photocopied.

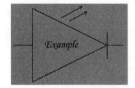

Example

Using these conventions, you would write 15 000 Ω as 15k and 2 200 000 Ω as 2M2. A resistance of 1.5 Ω would be written as 1R5, and a resistance of 0.47 Ω would be written as 0R47.

Resistors can be manufactured to practically any value that you want, but in practice there is no point in having such a vast range. A standard set of *preferred values* is used, and this also fits in with the manufacturing tolerances for resistors and other components. For example, the standard set values of 1.0 and 1.5 allow manufacturing with 20% tolerance with no rejected resistors. This is because if you take any pair of values on the scale, then a value which is 20% high for one value will overlap the amount which is 20% low for the next value. For example, 20% up on 1.0 is 1.0 + 0.20 = 1.20 and 20% down on 1.5 is 1.5 − 0.3 = 1.2, so that these values can overlap – a resistor of value 1R2 could be a 1R0 which was on the high side or a 1R5 which was on the low side. We can pick numbers from the standard set (for 5% tolerance) to suit 10% or 20% tolerances, as Table 2.1 shows. Resistors with 20% tolerance are not used to any great extent nowadays.

The preferred value numbers need only be in the range shown here, because we can multiply or divide by ten to obtain other ranges. For example, the number 4.7 can be used in the form 4R7 for 4.7 ohms, or as 0R47 (0.47 ohms), 47R, 470R, 4k7, 47k, 470k, 4M7, 47M and so on.

Summary

Resistors are manufactured using a range of preferred values that ensures there will be no rejects. Tolerances of 5% and 10% are commonly used. Values are written using the letter R in place of the ohm sign or the decimal point, and using k (kilo) to mean thousand and M (Mega) to mean million. This allows value to be specified without the need to write a large number of zeros.

$$\text{power} = V \times I$$
$$= I^2 R$$
$$= \frac{V^2}{R}$$

alternative
symbol

Figure 2.1 Power dissipated by a resistor. There are three versions of the formula so that you can use whichever is most suitable. For example, if you know values of I and R, use the I^2R formula.

A resistor behaves in the same way to all voltages and currents, whether these are steady voltages or currents; or waveforms. In other words, the relationship $V = RI$ is always true for a resistor which is kept at a constant temperature (Ohm's law). When a current passes through a resistor there will be a voltage across the resistor and power is converted into heat, Figure 2.1. Just as the $V = RI$ relation can be written in three ways, the power equations also exist in three forms.

Wherever there is resistance in a circuit, electric power is converted into heat and this represents waste, loss of energy. In addition, the conversion of electrical energy into heat means that the temperature of a resistor will rise when it is passing current. Unless the resistor can pass this heat to the air it will overheat and be damaged. Figure 2.1 shows the relationship between voltage, current, resistance and lost power, and also shows the symbols that are used to represent a resistor on a circuit diagram. Both the block and the zigzag symbol are in use, but outside the USA, the block is the preferred symbol nowadays.

This power loss in the form of heat is called *dissipation*, and is always associated with resistance (whether of resistors or in other components). If the dissipation is large, some method has to be used to keep the components from overheating, and this usually takes the form of cooling fins so that the heat can be more efficiently transferred to the air. There are some components (such as capacitors, see later) which do not dissipate heat, but any circuit will inevitably contain some resistors from which heat will spread to other components. Heat dissipation is particularly important in a circuit that contains active components as we shall see in Chapter 3.

When a resistor carries a steady current, there will be a steady voltage across the resistor ($V = RI$); and when there is a steady voltage placed across a resistor, there will be a steady current ($I = V/R$). The relationship applies also to waves, so that if a waveform of current passes through a resistor there will be a voltage waveform across the resistor whose value can be found from $V = RI$. When a resistor is used to convert a current wave into a voltage wave, we call it a *load resistor*. Load resistors are used along with active components.

When we use these $V = RI$ relationships with alternating currents and voltages, we normally use peak values for both voltage and current. Another option is to use RMS values (see later) for both quantities. Whatever type of measurement you use must be used consistently.

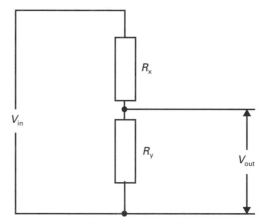

Figure 2.2 The potentiometer or voltage divider circuit. The output voltage is a fraction of the input which can be calculated if the resistor values are known.

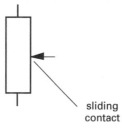

sliding
contact

Figure 2.3 The variable potentiometer symbol. This component is used to provide an adjustable voltage division and a typical use is as a volume control.

Resistors can be used to reduce the amount (amplitude) of a signal. Suppose, for example, that we connect two resistors in series (one connected to the end of another) as shown in Figure 2.2. If a signal voltage is connected across both resistors, as illustrated, then the output across one single resistor is a smaller signal voltage, and the size of this signal can be calculated from the sizes of the resistors. As a formula, this is:

$$V_{out} = V_{in} \times R_y/(R_y + R_x)$$

and it allows us to adjust signals to whatever amplitude we want to use. Suppose, for example that R_x is 10k and R_y is 15k, and we have a 20 V signal at the input. Since the total resistance is 25k, the output is 20 × 10/25, which is 8 V.

The dot that is placed where lines join is a way of emphasising that these lines are electrically connected. When you see lines crossing on a modern circuit diagram, this means that there is no connection between the lines.

This combination of two resistors is called a *potentiometer* or an *attenuator*, and we can manufacture an adjustable (variable) potentiometer which allows us to vary the values of both resistors. The symbol is shown in Figure 2.3,

Figure 2.4 The capacitor symbol. The basic symbol is this parallel-plate type (a) that indicates the nature of a capacitor as a pair of conductors separated by an insulator. (b) The symbol for an electrolytic capacitor, used for large capacitance values.

and you can think of it as a resistor with an extra contact which can be moved in either direction. This allows the output voltage to be adjusted (by altering the position of the contact) from the maximum (which is the same as the input) to zero. The potentiometer is used, for example, as a volume control in a radio.

The resistors in Figure 2.3 are connected in series, meaning that the current must pass through both resistors equally, one after the other. The alternative is parallel connection, in which the current splits between the resistors. When components are connected in parallel, the voltage across them is the same. Examples of series and of parallel connection will be pointed out in the course of this book.

The main use of resistors is to control current or as load resistors to convert current waves into voltage waves. The relationships $V = RI$, $I = V/R$ and $R = V/I$ hold for either steady or alternating voltages and currents. Two resistors in series can be used to reduce (*attenuate*) a signal voltage, and a variable version of this arrangement is a potentiometer, used to adjust signal levels.

Capacitors

A capacitor is a gap in a circuit, a sandwich of insulating material between two conductors which has capacitance (see later). The symbol for a capacitor, Figure 2.4, shows it as two conducting plates with a gap between them, and this can be used as a way of manufacturing capacitors, though we more usually find a solid insulator between the plates. Two exceptions are variable capacitors and electrolytic capacitors. Variable capacitors use two sets of plates that mesh with each other, and because one set of plates is carried on a spindle, turning the spindle will alter how much the plates mesh, and so alter the (small) capacitance between them. This has been used widely in the past for tuning radios. The electrolytic capacitor, by contrast, uses an acid jelly between metal plates, and the insulator is hydrogen gas that is generated by chemical action. This type of capacitor is used when a very large value of capacitance is needed in a small volume.

As far as steady voltages or currents are concerned, the capacitor is just a gap, a break in a circuit so that no steady current can flow. When you place a steady voltage across a capacitor there is no steady current. It's a different matter when an alternating voltage is used. When you move electrons on to one plate of a capacitor, the same number of electrons will leave the other plate, so that when you move electrons alternately to and from one plate, the same waveform will occur on the other plate, just as if it had been connected through a circuit. You can measure the alternating voltage and the alternating current and find the quantity which is V/I. This quantity is called capacitive *reactance*, and given the symbol X_C.

Unlike a resistor, a capacitor does not have a fixed value of reactance, because if you change the frequency of the supply wave, the reactance of a capacitor will change. When frequency increases, reactance decreases, and when frequency is decreased, reactance increases. There is, however, a quantity called *capacitance* which depends on the measurements of the capacitor and the type of insulating material. The value of capacitive reactance can be calculated if you know values for capacitance and for the frequency of the alternating voltage.

The natural unit for capacitance is the farad (named after Michael Faraday), but this unit is too large for the sizes of capacitors that we use for electronics circuits. We therefore use the smaller units of microfarad (μF), nanofarad (nF) and picofarad (pF). A microfarad is one millionth of a farad, the nanofarad is one thousandth of a microfarad, and the picofarad is one millionth of a microfarad. For example, a variable capacitor might have a maximum value of 300 pF; an electrolytic might have a value of 5000 μF.

A capacitor consists of an insulator between conducting plates and does not pass steady current. Alternating voltages will cause an alternating current to pass through a capacitor, and the ratio of V/I is constant if the frequency is not changed. This ratio is called capacitive reactance. A more fundamental quantity, called capacitance, can be calculated, in units of farads, from the dimensions of a capacitor and the type of insulator, and this quantity is a constant for a capacitor. The reactance at any frequency can be calculated from the capacitance value.

When an alternating voltage is applied to a capacitor and alternating current flows, the voltage wave is not in step with the current wave, but one quarter of a wave later, Figure 2.5. Contrast this with the behaviour of a resistor, which is also illustrated in this diagram. Because the maximum current through a capacitor happens at the time of zero voltage, and the maximum voltage at the time of zero current, there is no power dissipation from a perfect capacitor (one which has no resistance). Nothing's perfect, but capacitors get pretty close in this respect and their dissipation is usually almost unmeasurably small.

The amount by which voltage and current are out of step is expressed as a *phase angle*. If you think of a cycle of a wave as being caused by a coil rotating between the poles of a magnet, one complete wave corresponds to one complete turn of the coil, turning through 360°. On this basis, half a wave corresponds to 180° and quarter of a wave to 90°. We say, then, that the capacitor causes a 90° phase shift between voltage and current for an alternating supply, with the current wave ahead of the voltage wave. We say that in the capacitor, current *leads* voltage or, alternatively, that voltage *lags* current.

This idea of phase is very important in all branches of electronics, and you will need to recall it when we look at stereo radio broadcasting and colour TV principles. Phase is a quantity which we can alter for a wave just as we can alter frequency or amplitude.

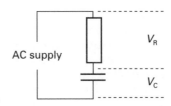

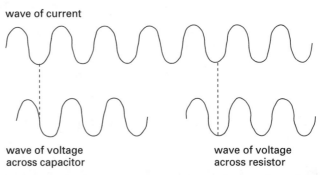

Figure 2.5 Phase angle. The phase of voltage across the capacitor is 90° later than the phase of current (which is also the phase of the voltage across the resistor).

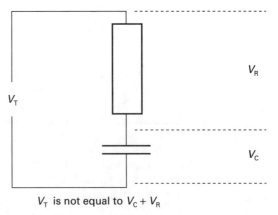

Figure 2.6 One effect of phase shift is that the total voltage across a series *R* and *C* circuit is *not* equal to the sum of the separate voltages across the components.

When a circuit contains a capacitor the wave of voltage in that complete circuit will not be in step of the wave of current. Suppose, for example, that a circuit contains both a capacitor and a resistor. The wave of voltage across the resistor will be in step with the current wave, but the wave of voltage across the capacitor will not be in step with the voltage wave across the capacitor. The total voltage across the circuit cannot be calculated simply. For example, in the circuit of Figure 2.6, the total voltage V_T is not equal to $V_C + V_R$. There are ways of calculating this, but they are a long way from simple addition, and this book is not concerned with mathematics.

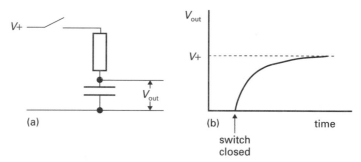

Figure 2.7 Charging a capacitor through a resistor. In this circuit (a), the voltage across the capacitor rises with a rate that is not constant, giving a curved graph (b). This is an *exponential* increase.

Alternating current through a capacitor is one quarter of a wave, 90°, ahead of the wave of voltage. This has two effects. One is that there is no dissipation in a capacitor except from the resistance of its conductors. The other is that the presence of a capacitor in a circuit causes the waves of voltage and of current in the whole circuit to be out of step.

Charge and discharge

Capacitors act as insulators for steady voltages, and as reactances for waves, but their behaviour with pulses is important. The capacitor can store a tiny amount of electrical charge when it is connected to a steady voltage, but if the connection is made (as it almost always is) through a resistance, then the charging action takes time, even if this time is measured in microseconds, and the time is not a simple measurable quantity.

Look, for example, at what happens in the circuit shown in Figure 2.7(a). This shows a voltage supply with a switch, a resistor and a capacitor. When the switch is open, disconnecting the circuit, there is no voltage across the capacitor. When the switch is closed, current flows and it will charge the capacitor, but as the voltage across the capacitor increases the amount of current is reduced (if you think in terms of water it has a shorter distance to fall), so that the rate of charging slows down.

The result is that a graph of voltage across the capacitor plotted against time takes the form shown in Figure 2.7(b). This is the type of graph shape that is called an *exponential rise*, and what makes it interesting is that it is never totally complete. Mathematically, the graph is described using a universal constant called 'e', the exponential constant which also appears in calculations of compound interest or of decay of radioactivity. A convenient rule of thumb makes use of what is called a time constant (*TC*).

The time constant of a combination of a capacitor and a resistor is the value of capacitance multiplied by the value of resistance. If the capacitance value is in units of farads and the resistance is in units of ohms, the time constant $R \times C$ is in units of seconds. More practical units are kΩ for the resistor and nF for the capacitor, giving time in µs. For example, using a capacitor of 20 nF and a resistor of 100k gives a time constant of 2000 µs, which is 2 ms (milliseconds).

The importance of the time constant is that we can take it that charging or discharging is over for all *practical* purposes (meaning about 95%) after a

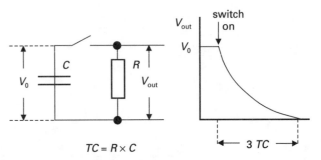

Figure 2.8 Discharging a capacitor through a resistor. The graph shows an exponential decrease, and we can take it as being complete in three time constants.

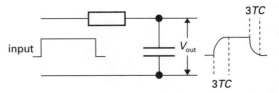

Figure 2.9 The effect of this type of circuit on a square wave, showing the time in terms of time constant *TC*. This is an integrating circuit.

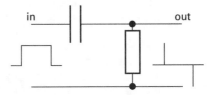

Figure 2.10 The differentiating form of the circuit. Remember that the voltage across the capacitor cannot change instantly, so that when the voltage changes suddenly on one plate, it must make the same change on the other plate, after which charging or discharging will alter the voltage.

time of three time constants. For more precision, a value of four time constants can be used. This makes it easy to work out the time of the waveform that is produced when a resistor and capacitor are used in a charging circuit. The charging shape is matched by a discharging shape. Suppose that a resistor is connected across a charged capacitor, as in Figure 2.8. The shape of the voltage/time graph is then as shown in the drawing, once again taking the process as being complete after three time constants.

For example, if the input to the circuit of Figure 2.9 is a square wave whose flat portion takes more than three time constants, we can draw the output waveform fairly easily. The first part is a charging curve taking three time constants, and the last part is a discharging portion that also takes three time constants. If the top of the square wave had a duration of more than six time constants, the remainder is unaffected. The effect is one that we noted in Chapter 1, of integration of the square wave.

Figure 2.10 shows a slightly altered circuit in which the components are rearranged and the output voltage is across the resistor. Now when there is a sudden rise of voltage at the input, the capacitor has no time to charge,

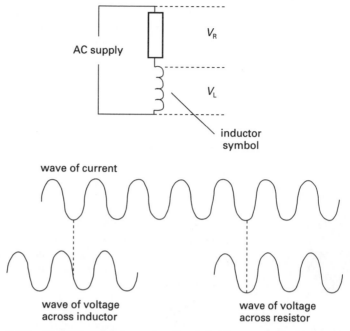

Figure 2.11 Phase angles in a circuit containing an inductor and a resistor. The phase of voltage across the inductor is 90° ahead of the phase of current (which is also the phase of the voltage across the resistor).

and the voltage across it is zero, which means that all of the input voltage appears across the resistor. Then as the capacitor charges, the voltage across the resistor drops to zero in the time of three time constants. If the input is then suddenly taken back to zero, the process repeats, with the voltage across the resistor dropping (so that the capacitor maintains its change), and then reducing to zero in three time constants. This is the action of a differentiating circuit.

Inductors

An *inductor* is a coil of wire. Since this wire has resistance, an inductor will pass steady current when there is a steady voltage across it, but the fact that the wire is wound into a coil makes it behave as more than just a resistance for alternating current. If the resistance is low, we find that the alternating current through the coil lags almost 90° behind the alternating voltage across the coil, as illustrated in Figure 2.11, which also shows the symbol for an inductor.

For any inductor a reactance can be measured, an *inductive reactance*. This quantity is constant only for a fixed frequency. If you increase the frequency of the alternating voltage, the reactance of the inductor also rises.

Once again, there is a quantity, called *inductance*, that can be calculated from the dimensions of the coil, and this quantity is constant for a coil. The more turns there are on the coil, the greater the inductance, and the inductance is also (greatly) increased when the coil is wound on a magnetic material (a *core*). Inductance is measured in units called henries (H), named after the US pioneer Joseph Henry. We often use the small units of millihenries (mH) and microhenries (μH), and the abbreviation for an inductor is L.

Inductors are not perfect – they have resistance because they are made from wire, so that there is always some dissipation from a coil, and the phase shift is never exactly 90°. Most of the coils that we use, however, have much larger values of reactance than resistance, and the imperfections are not too important. Modern electronic circuits avoid the use of inductors as far as possible, but when they are used, their symbol reminds you that an inductor is a coil.

Some applications call for very small inductance values. One example is a resonant circuit (see later) for a television UHF tuner or for a satellite receiver. To make the very small values of inductance the wire does not need to be coiled, and it is more usual to see a flat strip of metal used in place of a coil.

Summary

Alternating current through an inductor is one quarter of a wave, 90°, behind the wave of voltage. The dissipation in an inductor is only the amount you would expect from the resistance of its conductors. The presence of an inductor in a circuit causes the waves of voltage and of current in the whole circuit to be out of step.

Transformers

The simplest transformer consists of two coils wound over the same magnetic core. This has no effect as far as steady voltages are concerned, but for an alternating voltage the effect is very useful. When an alternating voltage is applied to one of the two coils, called the *primary*, there will also be an alternating voltage across the other coil, the *secondary*. The ratio of these voltages is the same as the ratio of the number of turns in the coils. For example, if the secondary coil has half the number of turns that the primary coil has, then the voltage across it will be half of the voltage across the primary. This arrangement is a *step-down* transformer, and you can just as easily make a *step-up* transformer, for which the voltage across the secondary coil is greater than the voltage across the primary. Figure 2.12 shows the circuit symbol and an elementary form of construction for a small transformer.

There is no gain of power in a transformer, it is a passive component. If you have 100 V across the primary coil and 1 A flowing, and a secondary that gives 200 V, then the secondary current cannot exceed 0.5 A, assuming no resistance and therefore no losses. We come close to this perfection in very large transformers, such as are used on the National Grid, but not on small transformers unless they are being used at frequencies higher than the normal 50 Hz of the power mains.

The transformer was invented by Michael Faraday, and it is the main reason for our use of AC for electricity distribution. Transformers allow us to convert one alternating voltage into another with only very small losses (caused by the resistance of the wire in the coils). We can generate electricity at a voltage which is convenient, such as 25 kV (25 thousand volts), and convert this to 250 kV for transmission, because for the same power level, the current flowing in the cables will be one tenth of the current from the generator. The less the current, the lower the power dissipated.

The main use of transformers in electronics is in converting the AC mains supply to the low voltage steady voltage needed for electronics in the power supply unit (PSU). The transformer converts the 240 V of the mains into a suitable lower voltage, and other components then convert this low-voltage AC into low-voltage DC.

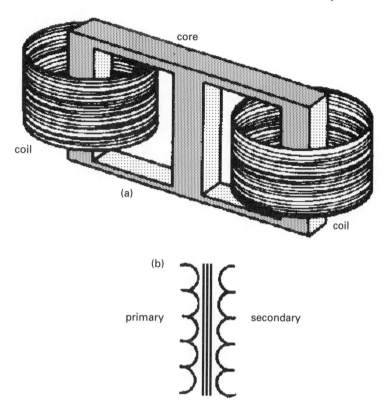

(a)

(b)

primary   secondary

Figure 2.12 The transformer (a) simplest practical arrangement, (b) symbol.

A coil of wire is simply a resistor as far as steady voltage is concerned, but for alternating voltages it behaves as an inductor. An inductor has inductive reactance, and causes a phase angle of almost 90° between current and voltage, with the voltage wave leading the current wave. The resistance of the wire means that an inductor is never perfect, but at the higher frequency ranges, the reactance can be very much greater than the resistance. The inductance of a coil can be calculated from its dimensions, and used to find reactance at any frequency. Reactance increases as frequency is increased. The inductance value is very much larger if the coil is wound on to a magnetic material, and when two coils are wound on the same core, the result is a transformer. A transformer can change voltage and current levels with almost no loss of power.

Resonance

We have seen that both capacitors and inductors affect the phase angle between current and voltage for AC. The effect, however, is in opposite directions, and a useful way to remember this is the word C-I-V-I-L. Say this as 'C – I before V; V before I for L' to remind you that for a capacitor (C) the current (I) wave comes before the voltage (V) wave, but the voltage wave comes before the current wave in an inductor (L). The US version of this is ELI ICE, using E for voltage.

Suppose a series circuit contains both capacitance and inductance along with the inevitable resistance, such as in Figure 2.13. How does such a circuit respond to alternating voltages? We can, of course, rule out any

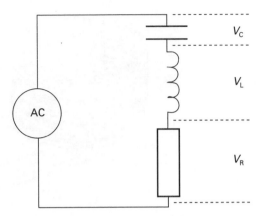

Figure 2.13 A circuit containing capacitance, inductance and resistance in series. This is a series resonant circuit.

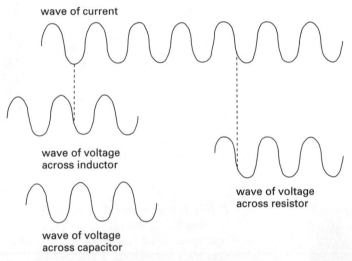

Figure 2.14 The voltages across the capacitor and across the inductor oppose each other. These voltages cancel each other exactly at the frequency of resonance.

possibility of steady current, because the capacitor will act like a break in the circuit as far as DC is concerned. The interesting thing is that if we pass an alternating current, the voltage across the capacitor will be in opposite phase, 180°, to the voltage across the inductor (because, compared to the current, the voltage across the inductor is 90° leading and the voltage across the capacitor is 90° lagging), Figure 2.14. The total voltage across the re-active components is the difference between the voltage across the inductor and the voltage across the capacitor.

This becomes particularly interesting when the reactance of the inductor is the same size as the reactance of the capacitor. When this is true, as it must be at some frequency, then the two voltages across the reactances will cancel out, and all that is left is the voltage across the resistor. This condition is called *resonance*, and in this series circuit the current is maximum at resonance. A graph of current plotted against frequency, near the frequency of resonance, looks as in Figure 2.15.

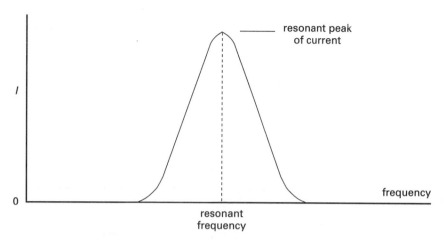

Figure 2.15 The current varies with the frequency for a series resonant circuit. This allows a particular frequency to be selected in, for example, a radio receiver.

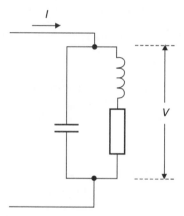

Figure 2.16 The parallel resonant circuit. The voltage across this circuit is a maximum at the resonant frequency.

A resonant circuit is one in which the effects of capacitance and inductance cancel each other for one particular frequency.

There is another way of connecting a capacitor and an inductor (with the inevitable resistance), which is illustrated in Figure 2.16. This is a *parallel* connection, and in this circuit, DC can pass because the coil is a wire connection. If we apply only an alternating current supply, however, we find that this time it is the *voltage* that becomes a maximum at the frequency of resonance when the reactances are equal in size.

A resonant circuit can act like a selective transformer, delivering an output which is at a much larger voltage or current than the input, for one frequency (in practice, a range or band of frequencies centred around one frequency). This is the effect that allows a radio to be *tuned* to one of a set of transmitting stations, using the selective transformer effect of the resonant circuit. Resonant circuits are also important for timing and for transmitting signals. A piece of quartz, made in the form of a capacitor that uses the

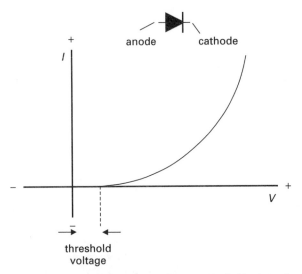

Figure 2.17   The diode and its graph of current plotted against voltage. Even in the forward conduction direction, there is no current flowing when the voltage is small (typically 0.4 V for diodes constructed from silicon). There is no current in the reverse direction unless a very large reverse voltage is applied, which will cause the diode to break down.

quartz as an insulator, will behave like a resonant circuit, with a step-up ratio much higher than can be achieved by any combination of inductor and capacitor, so that these quartz crystals are used to control the frequency of transmitters and also in clocks and watches.

When an inductor and a capacitor are used in the same circuit, their phase shifts are in opposite directions. When the sizes of the reactances are equal, the effects cancel so that, for alternating signals, the only effect is of resistance. For a series circuit, this causes the current to be a maximum at the resonant frequency and for a parallel circuit the voltage is a maximum at resonance. The resonance effect is used for selecting a frequency or a small range (a *narrow band*) of frequencies for purposes such as radio tuning. A quartz crystal can be made to resonate, and is more efficient than any inductor/capacitor combination, so that quartz crystals are widely used for timing and frequency setting.

Diodes

A diode is a passive component, but its construction follows the methods that are used for active components. A diode can be used with either steady or alternating supplies, but all resemblance to a resistor ends there, because a diode is not ohmic (*see* Chapter 1). A diode passes current in one direction only, and this is indicated by an arrowhead on the symbol that is used, Figure 2.17.

This illustration also shows a typical graph of current plotted against voltage. Unlike the corresponding graph for a resistor, this graph shows both positive and negative scales for current and voltage, because this allows us to show that the diode conducts in one direction only, and that it is not ohmic. The graph line is not straight even when the diode is conducting, so that there is no single figure of resistance that can be used – you cannot

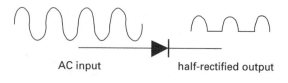

AC input   half-rectified output

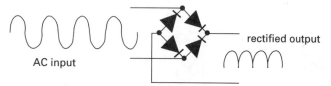

rectified output

AC input

Figure 2.18 Rectification of AC using diodes. The simple half-wave action is used for demodulation (see later), but the four-diode *bridge* circuit is almost universally used for AC to DC conversion for power supplies.

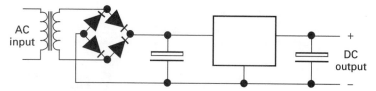

AC input   + DC output −

Figure 2.19 A stabilised power supply, using a transformer, a diode bridge and a stabiliser chip. The electrolytic capacitors ensure that enough charge is stored to maintain voltage for the short intervals when the output from the diodes approaches zero twice in each cycle.

specify a 3k3 diode, for example. The resistance is very high when the current is low, and becomes lower as current is increased. Even when the voltage across a diode is in the *forward* (conducting) direction, the current is undetectable until the voltage has reached a *threshold* level.

A diode will break down if a large enough reverse voltage is applied. Diodes can be manufactured whose *reverse breakdown voltage* is precise and stable, and these *Zener diodes* are used for providing a stable voltage level.

The effect of a diode on an alternating voltage supply is illustrated in Figure 2.18. The effect is like that of a commutator, allowing only half of the waveform to appear at the output. This effect is used in converting AC into DC, and also for a task called *demodulation* of radio waves, see Chapter 7. The circuit illustrated the simplest conversion circuit, half-wave rectification, and the much more common *full-wave bridge* circuit. A diode is a passive component, though the methods that are used to manufacture diodes are also used to manufacture active components.

Figure 2.19 shows a typical simple modern power supply unit (PSU) which uses a diode rectifier bridge circuit along with electrolytic capacitors and a voltage stabiliser IC. The transformer supplies AC at a suitable (low) voltage and the output of the transformer is connected to the input of the diode bridge circuit. The input to the stabiliser is a voltage that is higher than we need at the output, typically +18 V for a 12 V output. The stabiliser contains a voltage reference source, a diode which is operated with reverse voltage and which has broken down. Such a diode has a constant voltage across it even if the current varies, and it can be used in a circuit which compares this steady voltage with the output voltage of the chip, using this difference to control the output voltage.

The power supplies for computers and other circuits which use large currents (typically 20 A or more) at low voltages (typically 5 V or less) are constructed differently, using what is termed a *switch-mode* power supply. This uses an oscillator to supply pulses which are rectified and smoothed, and the output voltage is used to control the oscillator.

Chapter 3   Active components and ICs

The transistor

Modern active components are all based on transistors, which were invented in 1948. Before transistors came into use, the main active components for electronic circuits were vacuum valves, and these are still used for high-power transmitters. There are two main types of transistors, called respectively bipolar and field-effect transistors, and since the bipolar type came first into use, we'll start with that.

The transistor is a component that is created using a semiconductor crystal, and in a sense it is the inevitable result of the use of crystals in radio reception in the 1920s, because this started a line of research into crystal behaviour which led to the transistor, even though the materials are quite different.

This book is about electronics, not physics, so that the way in which the transistor works is beyond our scope, but we are interested in what the transistor is and what it does. The symbol that is used on circuit diagrams is a useful reminder, Figure 3.1. This shows three connections, labelled as *emitter*, *collector* and *base*, with an arrowhead on the emitter lead that points in the direction of current.

For the type of transistor that is illustrated in Figure 3.1, a steady voltage can be connected, with the positive end on the collector lead and the negative (earth) end on the emitter. No current passes, and the same would be true if the voltages were reversed, so that in this state the transistor behaves like an insulator.

With the collector positive and the emitter negative, if the base connection is now slowly made more positive than the emitter, a very small current will eventually pass between the base and the emitter. This pair of connections, emitter and base, in fact, is a diode. What is much more significant, however, is that when a small current passes between the base terminal and the emitter terminal, a much larger current will pass between the collector and the emitter. By much larger, we mean anything from 30 to 1000 times greater. The action is that a very small amount of current passing between base and emitter can control a much larger current passing between the collector and the emitter.

The name transistor comes from *transfer resistor*, the original name for the device.

Suppose now that we use a resistor connected to a steady voltage to pass a very small steady current through the base. This will, as we have seen, make a much larger current flow through the collector to the emitter. The transistor is acting as a steady current amplifier, making a large copy of a small current. If now we connect a signal waveform between the base and the emitter, this will cause the current in the base to fluctuate. The current in

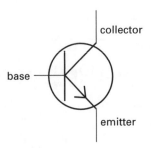

Figure 3.1   The circuit symbol for a transistor, in this case the NPN type which will pass current when both collector and base are at a positive voltage compared to the emitter.

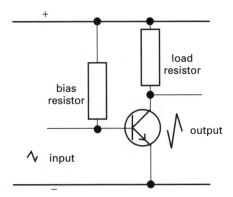

Figure 3.2   The simplest (and least satisfactory) form of transistor amplifier circuit, showing input and output waveforms. The output wave is inverted compared to the input and is of greater amplitude.

the collector circuit will also fluctuate, but with a much larger amplitude. For example, a current fluctuation of 1μA in the base circuit might cause a fluctuation of 1 mA in the collector circuit, a *current gain* of 1000 times. The transistor is now acting as a signal current *amplifier*. The steady current passing between base and emitter is called the base *bias* current. If we did not use this bias current, the base would not pass current until the input wave voltage reached about +0.6 V, so it could not have any amplifying effect on waves of small amplitude. With enough bias to allow current to flow, the transistor is a very effective amplifier for small signal inputs.

We can also make the transistor act as a voltage amplifier. The simplest (and least satisfactory) circuit is shown in Figure 3.2. A resistor is connected as a *load* between the collector and the positive supply terminal. A much larger value of resistor is connected between the base and the positive supply so that a small steady current flows through the base. Now if we add an alternating signal input at the base, taking care that the input is never so large that it causes the current to stop (by opposing the steady current), then the output will be a much larger voltage output.

This needs some thought. Imagine that the positive peak of the input wave has increased the base current. This will increase the collector current, and because there is more current through the load resistor the voltage across the resistor will be greater. Because there is more voltage across the resistor there is *less* between the collector and the emitter, so that the output

Figure 3.3 The more usual form of a single transistor amplifier. The use of additional resistors stabilises the steady currents that flow through the transistor.

wave is at its lowest, its negative peak. The output wave is a mirror-image of the input, an inverted wave, as is indicated in the drawing. A more practical circuit is illustrated in Figure 3.3.

A transistor of the bipolar type has three terminals, base, emitter and collector. A small current passing between base and emitter will cause a large current to pass between collector and emitter, with both base and collector positive in this example. This current gain effect can be used to construct a voltage amplifier in which the output is an amplified and inverted version of the input.

This principle of using a transistor and a load is at the heart of most of the electronic circuits (called *analogue* or *linear* circuits) that we were familiar with before digital circuits appeared. Transistors are even better suited to digital circuits, as we shall see in Chapter 10. The snag about using transistors for amplification is that the output is never a perfect copy of the input (though the imperfections can be made to be very small). A graph of the signal voltage output against the signal voltage input is not a straight line so that we say this is not a linear amplifier. There are ways of improving this situation, and we'll look at these in the following chapter.

NPN and PNP

Bipolar transistors can come in two forms called NPN and PNP. These P and N letterings indicate the type of carrier that takes most of the current in each region, so that the NPN transistor conducts mainly by electrons in the collector and emitter and mainly by using holes in the base region. The practical effect is that the NPN transistor is used with a positive supply to both the base and the collector; the PNP transistor is used with a negative supply to both base and collector. Some circuits use a combination of PNP and NPN transistors.

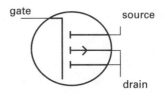

Figure 3.4 The symbol for a MOSFET, in this case the type called a p-channel MOSFET.

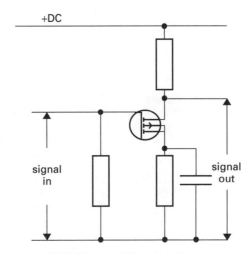

Figure 3.5 A typical MOSFET amplifier circuit.

Field-effect transistors

Field-effect transistors (FETs) have been available for almost as long a time as the bipolar type, but they were not extensively used until later. Nowadays, the field-effect type is used to a greater extent (in ICs, see later) than the bipolar type, and the most important field effect type is the metal-oxide-semiconductor FET, abbreviated (mercifully) to MOSFET.

The MOSFET uses quite different principles. Current can pass between two terminals called the *source* and *drain*, and this current is controlled by the *voltage* on a third terminal, the *gate*. On a circuit diagram, the MOSFET appears as in Figure 3.4, and the important point about it is that there is no current passing to or from the gate. This type of transistor needs a voltage signal only, not a current signal, so that the power needed at the input is very much smaller than is needed for a bipolar transistor, almost negligible. Depending on the design of the MOSFET, a bias voltage can be used or the MOSFET operated without bias. A typical amplifier circuit is illustrated in Figure 3.5.

MOSFETs do not provide as much amplification as bipolar transistors, and their main uses are not for amplifiers but in digital IC circuits, see later.

The MOSFET is a form of field-effect transistor which has become the most commonly used type of transistor. There are three terminals called source, gate and drain, with the voltage on the gate controlling the current between the source and the drain. An amplifier circuit uses a load resistor connected to the drain, and a bias supply to the gate if needed.

Switching

The uses of transistors, bipolar or MOSFET, for amplifier circuits require the designer to set correct bias, and use circuits that will provide the nearest to a straight-line graph of output plotted against input. These amplifier circuits all cause the transistor to dissipate power, because the optimum bias is usually with the collector or drain voltage set to about half of the steady voltage supply, and passing a current that is about the amount that needs to be supplied. For example, if the DC supply is at 12 V and a current of 50 mA needs to be supplied, then the transistor will be biased to a collector or drain voltage of 6 V and a current of about 50 mA. This corresponds to a power dissipation of 6×50 mW = 300 mW. This is typically the maximum that a small transistor can dissipate without overheating.

All linear amplifier circuits suffer from this problem, and when higher power outputs are needed transistors have to be designed so that they can release their heat through a metal casing or a metal stud to metal fins that will dissipate the heat into the air. For a few specialised purposes, water cooling can be used, but air cooling is much more common.

There is, however, another way of operating transistors, though it is a long way removed from their use as amplifiers. If a transistor of any type is operated without bias, it will not pass current when there is no input signal. With no current flowing, there is no power dissipation and so no heat to worry about. If a signal at the input (base or gate) now makes the transistor conduct fully, a comparatively large current will pass, but the voltage at the collector or drain will be low (as low as 0.2 V for some types), and the power dissipation will be small.

If we use transistors for working with pulses, then the power dissipation will be very low, particularly if the transistor is switched on for only a short time in each cycle. This switching use of transistors is not suited to linear amplifiers, but it is ideal for digital circuits, as Chapter 10 explains.

Transistor circuits can also be devised that make dissipation lower in switching circuits. For example, we have assumed so far that the load for a switching transistor will be a resistor, so that power will be dissipated in the resistor whenever current flows. If the load resistor is replaced by another transistor in a switching circuit, the 'load' transistor can be switched as well, ensuring that its dissipation is also low.

Transistors that are used for the amplification of waves (as distinct from pulses) are not as linear as we would like, and they dissipate power, risking damage to the transistors. The alternative way of using transistors is as switches, working with pulses that turn the transistor on for brief periods, so that linearity is not important and dissipation can be very low. This method of operation, however, is suited only to digital circuits.

Integrated circuits

Only a few years after the invention of the transistor, G. R. Dummer, working at a government laboratory, suggested that the processes used to manufacture individual transistors could be used also to manufacture resistors, capacitors and connections on the same piece of semiconductor material, making it possible to produce a complete circuit in one set of operations. Though the mandarins of the Civil Service who read his report attached no significance to it, others did. The USA was desperately trying to improve the reliability of electronic equipment for space use, and Dummer's suggestion was the answer to the problem.

It is possible to make individual electronic components reliable, but the weak point in large and elaborate circuits is the number of connections that

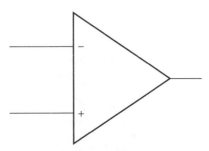

Figure 3.6   The symbol for an operational amplifier. The + and −
markings refer to the phase of input signals, not to DC supplies. The DC
supplies are usually balanced, typically +12 V and −12 V.

have to be made between components. If you can replace 20 individual
components by a single component then, other things being equal, you have
made the circuit 20 times more reliable, and this is what creating a complete
circuit in one set of operations amounts to. In addition, there is nothing that
restricts the idea to just 20 components in a circuit, though even the pioneers could not have imagined creating circuits with several million components on one small piece (*chip*) of silicon. This device is the *integrated circuit*
or IC. As is the case with most British ideas, we ended up buying into the
technology that we had failed to develop ourselves.

The space race between the USA and the USSR spurred the development
of ICs, so that by the 1970s we could buy complete circuits on a single tiny
chip of silicon that were vastly more reliable than anything we had ever
imagined. These ICs were mainly for digital circuits, though there are many
types of linear (amplifier) ICs also. In addition to reliability, the advantages
of these ICs include low cost, small size, low dissipation and predictable
performance. By this point in the twentieth century, making connections
between transistors is an almost forgotten skill used only where circuits are
created for special purposes.

The IC was a British invention, but you could be pardoned for not knowing
that. The principle is to improve reliability by using the manufacturing
methods for transistors to make resistors, capacitors and connections on
the same tiny chip of silicon as is used to make transistors. These chips
are mass-produced, and the elimination of external connections makes the
reliability of a circuit with a large number of individual components as good
as the reliability of a single component. In addition, the cost can be low
because of mass production, and the size of a complete circuit is little more
than that of a single transistor.

Linear ICs

Digital circuits have always accounted for a majority of the ICs that have
been manufactured, but there has always been one particularly important
type of linear IC, called the operational amplifier (or opamp). The circuit of
an opamp will be standardised, and only the designer is likely to know
exactly what has been done, so the user of an opamp works from a set of
figures that describe its performance. The usual symbol is illustrated in
Figure 3.6.

Instead of spending hours on the design of an amplifying circuit, a designer can use an opamp and (typically) two resistors to get the amount of

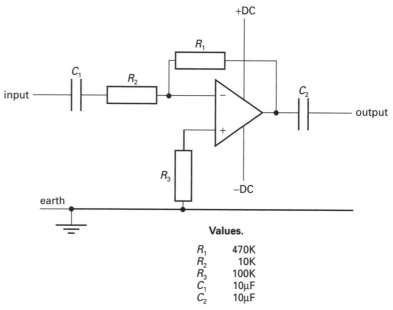

Values.

R_1	470K
R_2	10K
R_3	100K
C_1	10μF
C_2	10μF

Figure 3.7 A typical opamp circuit in which the gain (amount of amplification) is determined by the values of resistors R1 and R2.

gain that is required, and the calculations are simple. This does not mean that opamps can be used for any task that was formerly carried out by circuits using separate transistors (discrete circuits), but 99% of them is a reasonable estimate, sufficient to make the opamp a considerable boon.

The typical opamp has two inputs, marked + and –, and one output, and it is used with a balanced pair of DC supplies, typically +12 V and –12 V. The input markings do not refer to + or – supplies, but to the *phase* of signals. If you use the + input, the output signal will be in phase with the input signal. If you use the – input, the output signal will be in antiphase, a mirror-image waveform, like the output from a single-transistor amplifier. For a number of reasons, it is much more usual to take the input signal to the – input.

Figure 3.7 shows a typical amplifier circuit that will provide a gain of about 47 times. How do we know? Because that's the ratio of values of R_1/R_2. The opamp circuit is unknown, but the manufacturer guarantees that the gain of the chip is at least 100 000. The resistor R_1 is connected between the output and the input, so that some of the output is fed back, opposing the input. This reduces the gain, and it also makes the amount of gain depend only on the resistor values, not on anything inside the IC (unless some part of the IC circuit is overloaded). It's an end to complicated calculations and guesswork.

In this example, the DC power supplies have been shown. Opamp circuits are often shown with all power supply connections omitted. The resistor R_3 is used to ensure that the + input is connected to earth, and its value is not important.

Using an opamp is not an answer to everything, because the circuit might need to work at a frequency higher than the opamp can cope with – but a later type of opamp probably will. In addition, there are a host of special-ised opamps designed for specific purposes, and catering for almost all the exceptions to the general rules.

There is one serious handicap that affects all opamps. Because an opamp is designed for amplification, the transistors must use bias and therefore dissipate heat. This determines how complicated an opamp can be, in terms of the number of transistors in the circuit, because each transistor will con-tribute to the heat output and that heat must be transferred to the air if the opamp is not to overheat. The same applies to opamps that are intended to provide a power output to loudspeakers, electric motors and other devices.

The most common type of linear IC is the operational amplifier or opamp. This has a very high gain value, and is normally used in a circuit in which two external resistors control the gain. In the situations where such an opamp is not suitable, there are other designs that will deliver the perform-ance that is needed for more specialised purposes. The snag is that heat dissipation limits the complexity and power capabilities of an opamp.

Digital ICs

Digital ICs are the more common variety, mainly because of the vast number of digital devices (not just computers) that make use of these types of ICs. The transistors inside digital ICs are not being used as amplifiers, but as switches. This means that the heat dissipation for each transistor is very low, allowing digital ICs to be constructed using hundreds, thousands and even millions of transistors. In addition, heat-dissipating components (resistors) can be designed out because substituting a transistor for a resistor is easy when both use the same techniques (and a transistor can be physically smaller than a resistor).

Digital ICs deal with pulse inputs and outputs and use switching actions with very low dissipation.

The simplest digital ICs carry out just one type of switching action, and they perform the operations that are called logic actions, see Chapter 10. The type of circuits that can be constructed using these ICs would typically be used as controllers for machines, using several inputs to decide whether an output should be on or off. When should your washing machine start a cycle? Obviously, when the main switch is on, a program is selected, the drum contains some clothes, the water supply is switched on and the main door is closed. The machine must not switch on unless all of these 'inputs' are present, and this action of providing an output only for some particular set of inputs is typical of the type of circuits we call *combinational*. We'll come back to all that in Chapter 10. All of the first generation of digital ICs were intended to solve that type of problem, and these ICs are still in production more than 25 years on.

The next development was to make ICs that dealt with *sequential* actions, such as counting. These ICs required more transistors in each circuit, and as manufacturing methods improved, designers found that they could pro-duce not just the ICs that could be assembled into counters but the com-plete counters in IC form. At the same time, IC methods were being used to create displays, the LED and LCD displays that are so familiar now, so that all the components that were needed for a pocket calculator were being evolved together, and soon enough a complete calculator could be made using just one chip.

The pocket calculator story is a useful one to trace this part of the history of electronics. The first pocket calculators used several ICs, and they needed a considerable amount of assembly work. You could at that time buy DIY kits if you were curious to find out how the calculator was assembled, and such kits were also cheaper than a complete calculator. Nowadays, the calculator consists of just one IC, and there is practically no assembly. It costs more to assemble and package the components as a kit than to make and package the complete calculator, and costs are so low that the calculators can often be given away as a sales promotion.

Another thread of the story concerns the power required. The first pocket calculators needed four cells and gave about one month of use before these were exhausted. Power requirements have been so much reduced that some calculators are likely to be thrown away before the cell is exhausted, and it's possible to run calculators on the feeble power from a photocell (which converts light energy into electrical energy).

The early digital ICs were constructed using bipolar transistors, mainly because these were easier to construct at the time. Now the snag with bipolar transistors is that they need current inputs – no current flows between collector and emitter unless a current flows between base and emitter. The base current might be small, but it must exist, and so a bipolar transistor must dissipate more power than the MOSFET type, which needs no current between gate and source terminals.

Eventually, then, digital ICs started to be manufactured using MOSFET methods, and this allowed the number of transistors per IC to be dramatically increased. This packing of transistors is measured roughly by names that we use for *scale of integration*. This is described in terms of the number of simple logic circuits (gates) that can be packed into a chip, and the first ICs were small-scale integration (SSI) devices, meaning that they contained the equivalent of 3 to 30 logic circuits. The pace of development at that time (the 1960s) was very fast, so that the terms medium-scale integration (MSI), and large-scale integration (LSI) had to be introduced, corresponding to the ranges 30–300 and 300–3000 logic circuits respectively.

This is a good example of how technology always runs ahead of expectation. LSI soon became commonplace, and we had to start using very-large-scale integration (VLSI) for chips with more than 3000 gates per chip. Soon enough, chips containing 20 000 or more gates were being manufactured, but a new label, extra-large-scale integration (ELSI) was not introduced until more than one million gates could be put on a single chip.

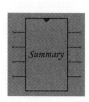

Digital ICs are classed in terms of the number of simple gate circuits that they replace on average. Modern chips are usually of the VLSI class, equivalent to 20 000 or more gates, and computer IC are often in the ELSI class, equivalent to one million or more gate circuits.

The micro-processor

The development of the microprocessor followed logically from the use of digital logic circuits, and this is an introduction only; more details follow in Chapter 12.

The microprocessor is a kind of universal logic circuit. Imagine you have a large number of digital logic ICs. You could make these into circuits, each of which would carry out some sort of control action, depending on how you connected the chips together. You might find that the circuits you used had a lot in common, and that you could make one circuit which used a number of switches, so that by setting switches you could change the overall action from one design to another. The next logical step is to make these switches in the form of transistors, so that the switching can be electrical. It is at this stage that the microprocessor becomes a possibility.

The first microprocessor was designed and built to a US military contract which called for a logic circuit with an action that could be controlled by inputs of signals called programming signals. The military contract was cancelled, and the manufacturer (Intel) was left with a large number of devices for which there was no known market. A few enthusiasts found that these chips could be used to make simple computers, and once this became known, a small firm called Altair put together a kit for building these computers. The sales of this kit were phenomenal, and by the end of a year most of the microcomputer firms that we know today were in business. It's ironic to note that larger firms such as IBM dismissed these machines as toys and predicted that they would all disappear within a year.

The microprocessor is a form of universal logic circuit with an action that can be set by the user. This allows the microprocessor to replace a vast number of other circuits and to be more flexible in use, because its action can be modified by programming it, using pulse inputs.

Cathode-ray tubes

The cathode-ray tube or CRT is an active component which nowadays is just about the only device you are likely to come across that uses the same principles as the old-style radio valves. The CRT converts an electrical signal into a light pattern, and though the principle was discovered in the 1890s the technology for mass production was not available until the 1930s and by that time was being pushed on by the needs of radar as much as by those of television.

The CRT can convert variations of voltage into visible patterns, and is used in instruments (oscilloscopes), for television, and for radar.

The three basic principles are as follows:

1 electrons can be released into a vacuum from very hot metals;
2 these electrons can be accelerated and their direction of movement controlled by using either a voltage between metal plates or a magnetic field from a coil that is carrying an electric current;
3 a beam of electrons striking some materials such as zinc sulphide will cause the material (called a phosphor) to glow, giving a spot of light as wide as the beam.

Nothing is ever as simple as basic principles might suggest. There are very few metals that will not melt long before they are hot enough to emit electrons, and at first only the metal tungsten was suitable. Tungsten was used for radio valves and for CRTs until a chance discovery that a mixture of calcium, barium and strontium oxides, comparatively cheap and readily available materials, would emit electrons at a much lower temperature, a dull-red heat. By the early 1930s, 'dull-emitter' valves were being mass-produced, and some CRTs were being manufactured for research purposes.

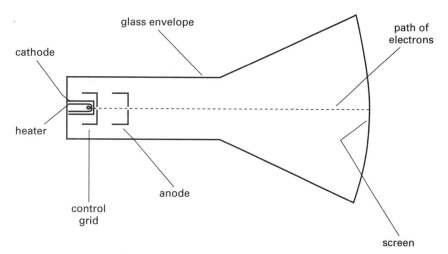

Figure 3.8 A very simple type of CRT which can produce an electron beam that in turn will make a spot of light appear on the screen.

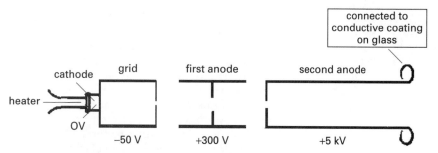

Figure 3.9 Typical voltages on the electrodes of a small CRT.

Figure 3.8 shows a cross-section of a very simple CRT. The cathode is a tube of metal, closed at one end and with that end coated with a material that emits electrons when it is red-hot. A coil of insulated wire, the heater, is used to heat the cathode to its working temperature. Because the far end of the tube contains conducting material at a high voltage (several kV), electrons will be attracted away from the cathode.

These electrons have to pass through a pinhole in a metal plate, the control grid. The movement of the electrons through this hole can be controlled by altering the voltage of the grid, and a typical voltage would be some 50 V *negative* compared to the cathode. At some value of negative grid voltage, the repelling effect of a negative voltage on electrons will be greater than the attraction of the large positive voltage at the far end of the tube, and no electrons will pass the grid – this is the condition we call *cut-off*.

Electrons that pass through the hole of the grid can be formed into a beam by using metal cylinders at a suitable voltage – Figure 3.9 shows a set of typical voltages for a small CRT. By adjusting the voltage on one of these cylinders, the *focus electrode*, the beam can be made to come to a small point at the far end of the tube. This end is the screen, and it is coated with a material (a *phosphor*) that will glow when it is struck by electrons. The phosphor is usually coated with a thin film of aluminium so that it can be connected to the final accelerating (*anode*) voltage. The whole tube is pumped

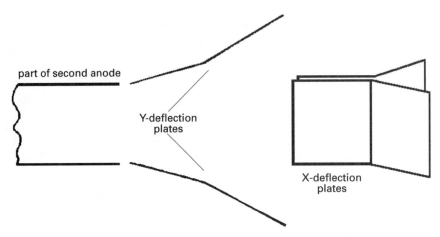

Figure 3.10 Using metal plates to deflect the electron beam. The plates are sloped and bent to ensure that the beam does not strike them.

to as good a vacuum as is possible; less than a millionth of the normal atmospheric pressure.

This arrangement will produce a point of light on the centre of the screen, and any useful CRT must use some method of moving the beam of electrons. For small CRTs, a set of four metal plates can be manufactured as part of the tube and these deflection plates will cause the beam to move if voltages are applied to them. The usual system is to arrange the plates at right angles, and use the plates in pairs, Figure 3.10, with one plate at a higher voltage and the other at a lower voltage compared to the voltage at the face of the tube. This system is called *electrostatic deflection*.

There is an alternative method for deflecting the electron beam which is used for larger tubes, particularly for computer monitors, radar and TV uses. A beam of electrons is a current flowing through a vacuum, and magnets will act on this current, deflecting the beam. The easiest way of doing this is to place coils around the neck of the tube and pass current through these coils to control the beam position on the face of the tube. This magnetic deflection method, Figure 3.11, is better suited for large CRTs such as are used for TV or radar.

The CRT will use either electrostatic or magnetic deflection, so that the beam of electrons can have its direction altered so that the light spot can appear anywhere on the face of the tube. Magnetic deflection is used for the large TV, computer monitor and radar tubes; electrostatic deflection for the smaller tubes for measuring instruments.

The form of deflection that is most common for CRTs is a *linear sweep*. This means that the beam is taken across the screen at a steady rate from one edge, and is returned very rapidly (an action called *flyback*) when it reaches the other edge. To generate such a linear sweep, a sawtooth waveform, Figure 3.12, is needed. An electrostatic tube can use a sawtooth voltage waveform applied to its deflection plates, and a magnetic-deflection can use a sawtooth current applied to its deflection coils. The difference is important, because the electrostatic deflection requires only a sawtooth voltage with negligible current flowing, but the magnetically deflected tube

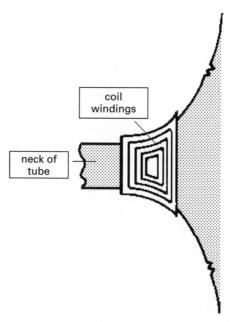

Figure 3.11 Using magnetic deflection, with coils wrapped around the neck of the tube. This form of deflection allows very short tubes to be constructed.

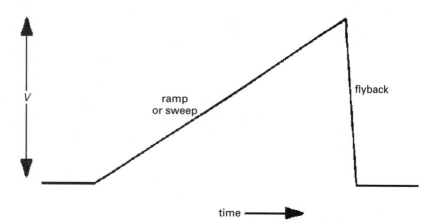

Figure 3.12 The ramp or sweep wave which is needed to scan a CRT. For an electrostatically deflected tube, this would be a voltage waveform applied to plates; for a magnetically deflected tube this is a current waveform applied to coils.

requires a sawtooth *current*, and the voltage across the deflection coils will not be a sawtooth, because the coils act as a differentiating circuit. In fact, the voltage waveform is a pulse, and this is used in TV receivers to generate a very high voltage for the CRT, see Chapter 8.

LCD screens

Vacuum CRTs are still the dominant technology for the larger TV screens and for measuring instruments, but smaller displays can now be made using LCD (liquid-crystal display) principles. The principle is that some liquid

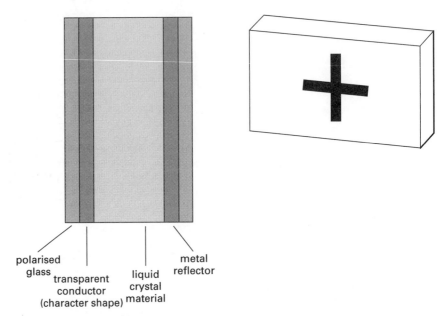

Figure 3.13 Simplified cross-section of a liquid-crystal display cell.

materials (including the cholesterol that clogs up your arteries) respond to the electric field between charged plates, and line up so that one end of each large molecule points towards the positive plate and the other end to the negative plate. In this condition, light passing through the material is polarised, so that if you look at the material through Polaroid sheet, the amount of light that passes alters when you rotate the Polaroid sheet.

An LCD display uses a set of parallel plates, of which one is a transparent conducting material, and one wall of the display is made of polarising material, Figure 3.13. With no voltages applied, light passes through the polarised front panel, through the liquid and is reflected from the metal backplate and out again. When a voltage is applied, the liquid between plates is polarised, and light can no longer be reflected back, making that part of the display look black. This type of display has been used for many years for calculators because it consumes very little power and yet gives a very clear output in good lighting conditions.

The shapes that are to be produced are determined by the shape of the film of transparent conducting material at the front of the cell. For example, if this film is deposited in the shape of a '2', the cell will show this digit when activated.

For use as a display for portable computers, each position on a screen must be capable of being individually controlled, making the construction of such a screen much more difficult than that of a calculator display. For example, the usual type of computer screen is required to be able to display 640×480 dots, a total of 307 200 dots – and that means 307 200 tiny LCD cells each of which can be set to black or clear. LCD displays of this type have to be combined with ICs to make the connections useable, and the most recent types are active, meaning that one or more integrated transistors will control each dot.

Though small screens can be constructed reliably, and colour displays can be constructed by using several layers with dye added to the liquids, the manufacturing of large screens is still too difficult and too costly to make LCD a competitor for the CRT for applications other than portable computers (where cost is not so important a factor) and for miniature TV receivers. At the normal rate of progress of the electronics industry, however, this cannot last for long, and LCD is likely to replace the CRT for all purposes, except in instruments, by the end of the twentieth century.

Chapter 4   Linear circuits

Linearity

There are two main types of electronic circuits, labelled as *linear* and as *digital* circuits. Though the two types of circuits can be constructed using the same set of components (both active and passive) the active components are used in very different ways and the waveforms that are processed are very different.

A linear circuit is designed to make a *scaled* copy of a waveform, and by scaled we mean that the amplitude of the output of the linear circuit is related to the amplitude of the input waveform. The output amplitude might be smaller, in which case we often call the circuit an *attenuator*. The output amplitude might be equal, in which case we call the circuit a *buffer*. Most commonly, the output amplitude is greater than the input amplitude, and the circuit is an *amplifier*. If the circuit is truly linear, the output waveform has the same frequency and the same waveshape as the input waveform – it is a true copy at a different scale, Figure 4.1, and the ratio of the output amplitude to the input amplitude is called the *gain*.

The name of *linear* circuit arises from drawing graphs of output amplitude plotted against input amplitude. For a perfect linear amplifier, this graph should be a straight line, hence the name linear. As it is, because of the imperfections of active components, amplifiers are never perfectly linear, though we can obtain almost perfect linearity in a buffer circuit, and perfect linearity in an attenuator which uses only passive components. Figure 4.2 shows perfection and some common varieties of imperfections in graphs for amplifiers.

A linear circuit is one for which a graph of output plotted against input is a straight line. Linear circuits are used in analogue designs, though not all analogue circuits need be perfectly linear.

The most common imperfection is curvature – the graph line is curved rather than straight. This means that the amount of gain, the scaling, alters as the input amplitude alters. For example, large-amplitude signals might be amplified more than small-amplitude signals. This type of non-linearity is sometimes required, and the human ear itself is non-linear, a way of protecting us from the noise of explosions, aircraft and discos. In general, though, for purposes such as sound reproduction we need amplifiers that are as linear as we can get them, and a target of 0.1% deviation from a straight-line graph is taken as a reasonable target for a hi-fi amplifier

Another imperfection that is noted in the drawing is limiting. When an amplifier limits, the output amplitude stays constant even though the input amplitude is changing. This causes the shape of the waveform at the output to be severely distorted, flattening at the tops as in Figure 4.3. All amplifiers will limit if the input amplitude is too large, so that a complete amplifier

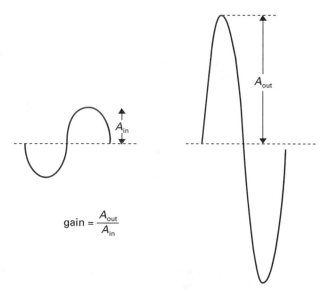

$$\text{gain} = \frac{A_\text{out}}{A_\text{in}}$$

Figure 4.1 Amplitude and gain. The amplitudes must be measured in the same way, and peak amplitude is illustrated here.

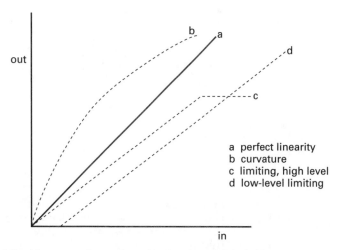

a perfect linearity
b curvature
c limiting, high level
d low-level limiting

Figure 4.2 Linear and non-linear behaviour of amplifiers.

system has to be designed so that limiting cannot occur even if the volume control is turned full up. If an amplifier is designed to have different input amplitudes, as it must if it is supplied from different sources such as tapeheads, pick-ups, CD players etc., then there must be adjustments provided so that the user can set each input to the same level so that the master volume control does not need to be adjusted when you switch from one source to another.

The third type of non-linear behaviour illustrated is another form of limiting. The amplifier simply does not amplify signals that are at a very low amplitude level. This type of non-linearity has in the past been used deliberately to reduce the noise signals from cassette tapes, but, like the other types it is undesirable in normal use. When transistor amplifiers first appeared, a form of this type of distortion, called *cross-over distortion*, was a

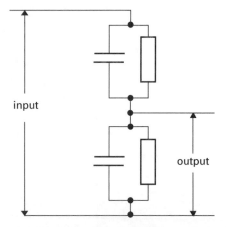

Figure 4.3 Squaring is the effect on the waveform of some type of limiting action in an amplifier.

Figure 4.4 An attenuator formed from passive components does not cause any non-linear distortions. The type illustrated here is a compensated attenuator which will work over a very wide range of frequencies.

common fault which delayed the acceptance of transistors for high-quality linear amplifiers until solutions were found.

Non-linear behaviour includes curvature and limiting. Though these effects are sometimes deliberately used, high-quality amplifier designers aim to reduce non-linearity to as low as can economically be attained. The amount of non-linearity is often expressed as the percentage by which the graph line deviates from a straight line, and a usual target for high-quality amplifiers is 0.1%, one part in 1000.

An attenuator that is constructed from passive components, such as the type shown in Figure 4.4 is perfectly linear for the normal signals that we are

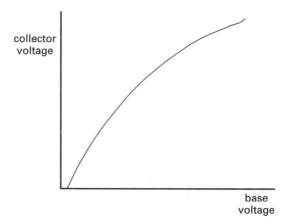

Figure 4.5 The non-linear behaviour of a transistor at its worst. Correct bias and choice of transistor type can help, but the graph shape is always a curve.

concerned with. This is because non-linearity is caused almost exclusively by active components. Passive components contribute to non-linearity only if they are overloaded, and that is reasonably easy to avoid.

The root cause of non-linearity in active circuits is that a simple plot of collector wave voltage against base wave voltage for a transistor in a simple amplifier circuit (Figure 4.5) is curved, and there is no collector output for small base waveforms. Careful attention to transistor biasing and circuit design can result in a graph shape that is closer to a straight line, eliminating the region where there is no collector output, but there is always some curvature. The problem is tackled by cunning circuit design that applies corrections to the output wave (a system called *negative feedback*), but this is not in itself a way of making a poorly-designed amplifier perfect; though it can make a good design work better.

A buffer circuit makes use of an active component circuit, but with the output amplitude the same as (or slightly less than) the input amplifier. For this type of action, it is possible to make transistors work in a fairly linear way so that distortion can be very small, though never zero. The purpose of a buffer is to prevent *loading* of a signal. For example, suppose that the input to an amplifier is a signal of 1 mV which can supply only 1 µA of signal current. Now if we connect this to an amplifier which will pass 10 µA at its input, we are asking too much of the signal source, loading it, and this is very likely to cause non-linear behaviour. If we place a buffer stage between the amplifier and the source, using a buffer which takes a current much less than 1 µA of current, then we avoid the loading and improve the performance of the whole amplifier. Buffers are used in all types of circuits, linear and digital, for this same purpose, to avoid taking more current from a signal source than it can comfortably supply.

The simplest type of transistor buffer circuit is illustrated in Figure 4.6. This is called an emitter-follower, and the output is in phase with the input. The amplitude of the output is less than the amplitude of the input, but the power available is more because the emitter current of the transistor can be much greater than the base current. This is an example of 100% negative feedback, because the input to the transistor is the voltage between base and emitter, but this is equal to the input signal (between base and earth) minus the output signal between emitter and earth.

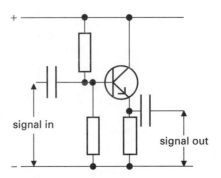

Figure 4.6 An emitter-follower circuit, a simple example of a buffer which provided power gain without voltage gain.

Buffers are linear circuits with zero gain, and it is fairly easy to ensure that they are almost perfectly linear. Buffers prevent too much current being taken from the source of a signal, and are used to isolate one section of a circuit from the next. The emitter-follower circuit is a simple type of buffer.

Gain

An amplifier carries out the action of making an enlarged copy of the wave-form that is used as its input signal, and the ratio of the output signal to the input signal is called the *gain* of the amplifier.

This could be written, for example, as ×10, ×100 or even ×1000, but we seldom use this way of expressing gain.

Voltage gain is defined as (output signal voltage)/(input signal voltage). We can also define power gain as (output power)/(input power).

Expressing gain as a simple ratio of voltages is often useful, particularly if we are aiming for some definite signal output level, but for amplifiers that deal with signals in sound or vision systems, the *decibel* scale is more useful. There's nothing mysterious about it, except that it is often misused. We have known for more than a century that human ears and eyes do not have a linear response. For example, when the amplitude of two sound waves is compared, one with twice the amplitude of the other, your ears do not hear one sound as twice as loud as the other. Similarly, two lights, one with twice the amplitude of the other, do not appear to your eyes so that one looks twice as bright as the other.

The type of scale that your senses use is logarithmic. These logarithm units of comparison are called decibels, named after Alexander Graham Bell who we remember for the invention of the telephone, but whose main interest in life was the problem of deafness – in fact, the first telephone was invented as a deaf-aid.

If one amplitude is 100 times the size of another, our senses tell us that it is something like 40 times, which is 20 times the logarithm of 100. As a formula this is $20 \log G_V$, where G_V is the simple voltage gain. If this is applied to power gains, it becomes $10 \log G_P$ where G_P is the simple power gain.

Strictly speaking, decibels should be used only for comparing power levels, using the 10 log G_p formula. The use of decibels for comparing voltages, however, is so common that it can't be ignored, though it is really valid only when the voltages are measured across the same value of resistance (which implies that the current has been amplified as much as the voltage).

The trap to watch for is that a decibel value always refers to a *ratio*. You cannot, for example say that a noise is at 100 dB unless you specify what your comparison is. 100 dB means that the noise is that much louder than something else, but you have to specify what that is. There are agreed comparisons for sound, but they are based on measured power rather than something subjective like a demented fly at 20 feet.

The gain of an amplifier is a ratio and is therefore always specified in decibels, and this form is also used for attenuators. For example, you might specify a –10 dB attenuation, meaning that the output voltage is only about 0.3 times the input voltage. The minus sign indicates that this is a ratio of input to output, not output to input, with the output amplitude less than the input amplitude.

Because our senses work on a logarithmic scale, it is often more convenient to express gain in this way, and this is the purpose of the decibel scale. For power gain, the number of decibels is equal to 10 log G_p, where G_p is the power gain, and this is the true definition of the number of decibels. For many purposes, however, it is convenient to work in voltage terms, and the number of decibels corresponding to a voltage gain is 20 log G_v, where G_v is the voltage gain.

Frequency response

The graph of output amplitude plotted against input amplitude shows up linear or non-linear behaviour, but there is another form of distortion that can affect any amplifier, and even passive devices such as attenuators. This is *frequency distortion*. Some amplifiers are intended to work with a single frequency or a narrow band of frequencies. These are tuned amplifiers and we shall look at them later in this chapter. More usually, an amplifier has to work for a range of frequencies, and it must have the same gain for all the frequencies in that range, which is its *bandwidth*.

For example, an amplifier that is used along with a CD player should be able to deal with the frequency range of about 30 Hz to 20 kHz, the *audio range*. These frequencies represent the limits of a (young) human ear, and as we get older our ability to hear the higher frequencies is steadily reduced – one estimate is that the rate is 1 Hz less each day, so that the upper limit can eventually be as low as 10 kHz or less. The ear is at its most sensitive at a frequency of around 330 Hz (the average frequency of the female voice), and is reduced at the very low as well as the very high ends of the frequency range. An amplifier designed to deal with this frequency range is an *audio amplifier*. This is the most common form of amplifier because every radio and TV receiver includes an audio amplifier stage.

A typical graph of gain plotted against frequency is illustrated in Figure 4.7. The gain is in dB units, and the frequency is shown on a logarithmic scale, so that the markings are 1, 10, 100, 1000 and so on, at equal intervals. This type of graph is called a *frequency response* graph.

A logarithmic scale is one that uses a fixed distance to represent a fixed multiple. For example, you might use 1 cm on the graph to represent each tenfold change in a quantity leading to a 1, 10, 100, 1000 scale with these

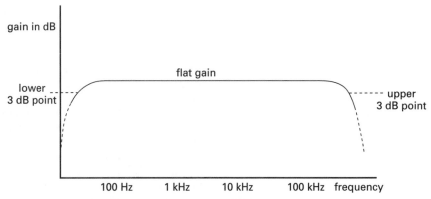

Figure 4.7 A frequency response graph, showing the 3 dB points, between which the bandwidth is measured.

numbers at equal intervals. On such a scale, the mark midway between 10 and 100 does not represent 50 (it is closer to 31).

If we used a linear scale, with the distance proportional to the frequency, the length of the frequency scale would be ridiculous. For example, if we allowed 1 cm for 10 Hz, the length of this scale would be about 2000 cm. In addition, because the decibel scale is logarithmic, it makes sense to use this type of scale also for frequency.

In the example, the frequency response is level for most of the range, with a downturn at each end. The convention is to take the frequency response as extending between the points where the level is 3 dB below the flat (maximum) portion. The basis of this is that 3 dB is an amount that is only just significant and noticeable to ears or eyes, so we can ignore variations of less than this amount. Even if the frequency response graph has no flat portion, we can ignore variations that are less than 3 dB. One of the advantages of using the decibel scale is that this amount of variation looks small on the graph – a 3 dB variation in terms of gain means a twofold change (twice or half) which would look very large on a linear graph.

The bandwidth of an amplifier is defined as the range of frequencies between which the change in gain is 3 dB. For audio amplifiers, this is usually quoted in terms of frequencies, such as 20 Hz to 20 kHz. For RF amplifiers (see later) this will be expressed as the difference between the frequencies, such as 5 kHz, 100 kHz, 5.5 MHz and so on.

No amplifier can have a perfect frequency response. The use of capacitors to carry signals between sections of a circuit limits the gain for the low frequencies, and there are also limits on gain at high frequencies caused by transistors themselves and by stray capacitance. Stray capacitance is capacitance between different parts of the circuit, and this arises because a capacitor is an insulator sandwiched between conductors, so that any two pieces of metal separated by air must have some capacitance. These strays filter off the highest frequencies and in some circuits the design must be able to compensate for such losses. In some circuits there may be hills and dales on the frequency response graph because of resonance effects.

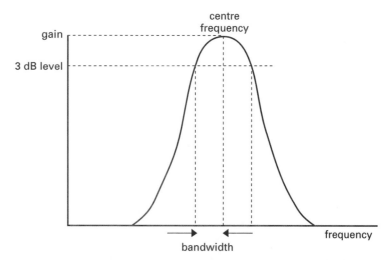

Figure 4.8 The bandwidth for a tuned circuit is measured to one of the 3 dB points on either side of the tuned (centre) frequency.

Broadband amplifiers

There is a special class of linear amplifiers called broadband amplifiers. This name is reserved for amplifiers which have a much wider frequency response than the ordinary audio amplifier. For example, the waveform that carries the picture information for a TV receiver is called the *video* signal. Ideally, an amplifier for this signal should have a bandwidth of 0 to 5.5 MHz (0 MHz means that changes in the steady voltage level are also amplified). In such a signal, there is a DC portion which carries the information on the overall brightness of the screen, while the highest frequencies carry the information in the finest detail.

This bandwidth is much greater than is used for audio signals, and to put it in perspective, it is more than five times as much as the whole of the medium-wave band on a radio. Even this, however, is small compared with some other bandwidths. A good computer monitor, for example will use a 30 MHz bandwidth, which is why we use (costly) monitors for computers rather than cheap TV receivers, and why computer images that look so clear on a good monitor look so fuzzy on a TV screen. Some measuring instruments that are used to find the amplitude and frequency of waveforms need amplifiers with a bandwidth of 50 MHz or more.

These broadband amplifiers need to be designed with great care, and the physical arrangement (*layout*) of the components is just as important as the theoretical design of the circuit. Even larger bandwidths are used in connection with the amplifiers that are used for receiving radio signals, particularly the amplifiers that are contained in satellite receiving dishes.

Tuned amplifiers

The broadband amplifiers that are used, for example, in a computer monitor are classed as *untuned* (or aperiodic) amplifiers, but we also make considerable use of tuned amplifiers. Recalling Chapter 2, a combination of an inductor and a capacitor is a tuned circuit that has a peak response at some frequency. This value of frequency, the tuned frequency, can be calculated from the amounts of inductance and capacitance, but the response of the circuit shows some gain for a range of frequencies around the tuned frequency. As usual, the bandwidth is calculated as the frequency range between the points where the response is 3 dB down (the –3 dB points), Figure 4.8.

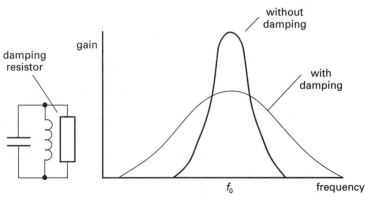

Figure 4.9 The effect of adding a damping resistor to a tuned circuit is to broaden the bandwidth and reduce the gain.

A tuned amplifier is one with a bandwidth that is centred around a single frequency. This is achieved using resonant circuits.

This bandwidth is important, because it affects the use that can be made of radio waves. As will be seen in Chapter 6, a radio transmission makes use of a band of frequencies that lie around a central maximum which is the frequency set at the transmitter. This range is, for medium-wave broadcasts, very small, only about 5 kHz on each side of the tuned frequency. This, in turn, means that the frequency range for sound on medium-wave radio is only about 5 kHz. As far as speech is concerned, this is adequate, but it is nothing like adequate for good reproduction of music. In addition, the older methods of using a radio wave to carry sound signals are very susceptible to interference both from natural (for example, lightning) and man-made (for example, car ignition systems) causes.

This problem has been around for a long time, and it was solved by an amazing genius called Edwin Armstrong (whose other inventions have also been landmarks in radio). In the 1930s, Armstrong developed a method, called *frequency modulation* (FM), for carrying high-quality sound on radio waves. This, however, requires a bandwidth of about 100 kHz, so that we need to be able to construct tuned amplifiers with this range of bandwidth for FM radio use. Television requires much larger bandwidths, and a typical TV broadcast might use a tuned frequency of 400 MHz, using a bandwidth of about 6 MHz.

The simple tuned circuit, noted in Chapter 2, cannot easily provide such large bandwidths, so a variety of circuit tricks have been used to broaden the bandwidth. All of these methods sacrifice gain so as to obtain more bandwidth, and the simplest is to connect resistors across tuned circuits. These are called *damping* resistors, and their action, as illustrated in the graph of Figure 4.9, is to reduce the gain of the tuned circuit but broaden the bandwidth. The reduction in gain can, if necessary, be overcome by increasing the number of tuned amplifying stages.

Another method of increasing bandwidth but maintaining gain is to use several tuned stages, but with each tuned to a different frequency around the central frequency. This is called *stagger tuning*, and it can provide the wide band amplification that is needed for the reception of TV signals (the broad-band amplifier for video is used at a later stage, see Chapter 8).

One important point about tuned amplifiers is that they need not be particularly linear, though they are classed as linear amplifiers. Any tuned amplifier uses a tuned circuit as a load, and even if the amplifier limits, the tuned circuit will complete the wave. This is because a tuned circuit acts like a pendulum which once set into motion keeps swinging. The tuned circuits are used as loads, and once a transistor has started to pass a wave of current through the tuned circuit, the voltage across the tuned circuit will be a complete wave even if the transistor stops passing current for part of the time.

Tuned amplifiers use resonant circuits tuned to a particular frequency, and the circuits can be arranged to give whatever bandwidth is required. Quite simple circuits can provide narrow bandwidths, but for wide bandwidths of several MHz more elaborate designs are needed.

Power and RMS

Using a linear voltage amplifier on a feeble signal will result in an output which is a signal that is at a much higher voltage level, but we cannot necessarily use this signal to operate devices like a loudspeaker or an electric motor. The reason is that these devices need a substantial amount of current passed through them, and a voltage amplifier cannot supply large currents. What we need is a *power amplifier*.

For example, a voltage amplifier might provide an output that was a voltage wave of 6 V which could supply no more than 1 mA of current. The power amplifier might need to be a wave with a voltage of 6 V (or slightly less) and a current of 2 A.

The name 'power amplifier' is misleading, because *any* amplification of voltage or current implies power amplification. In fact, the PA is a current amplifier, but since it is used to supply power to devices like loudspeakers, the name of power amplifier is more common.

Power amplifiers are used in applications that have no connection with loudspeakers. For example, the familiar dot-matrix printer for computers uses a set of power amplifiers to drive the pins in the printhead. There are usually 9 or 15 such pins, and each is moved by passing a current through a coil that surrounds the pin. This action requires power (so that the printhead dissipates a considerable amount of heat and can become very hot), so that a power amplifier is used to supply each coil.

The voltage gain of such an amplifier is often very low, sometimes less than unity (so that the output voltage is less than the input. The current waveform, however, has a much greater amplitude at the output, and current gains of 1000 or more are common. Because voltage gain is unimportant, it is easier to make a power amplifier linear for small signal amplitudes, but quite another matter to make it linear over the whole range of signals that it must cater for, particularly when the output is connected to something like a loudspeaker which behaves like a complicated circuit of resistors, capacitors and inductors.

$$V_p \times I_p = 2\,\text{W}$$

Figure 4.10 Signal power. For a wave, using peak values of V and I gives a power that is too large by a factor of two. We therefore use RMS values such as $V_p/\sqrt{2}$ and $I_p/\sqrt{2}$ which will give the correct amount when multiplied.

Summary

Power amplifiers are usually current amplifiers which are needed when a load, like a loudspeaker or electric motor, has to be supplied with a waveform that has been electronically generated.

The calculation of power for waveforms is another matter we need to look at. For steady voltages, the power dissipated in a resistor is easily calculated by multiplying the value of current through the resistor by the value of voltage across the resistor. In symbols, this is $V \times I$, and when units of volts and amps are used for voltage and current respectively, the power is in the units called watts (W), named after James Watt who first converted the power of heat into mechanical power in a steam engine.

What do we measure on a wave that can correspond to steady voltage and current? If we measure the peak wave value of voltage and current, Figure 4.10, we find that the power figure we get is twice as large as it would be if V and I were steady values. In symbols, $V_p\, I_p = 2$ W. This is not exactly surprising, because it's obvious that multiplying the peak value of these varying values must give a result that is less than one would obtain by multiplying steady values.

This is equivalent to using V and I quantities that are $V_p/\sqrt{2}$ and $I_p/\sqrt{2}$ respectively – if you multiply these quantities together you get $V_p I_p/2$, which is the true power in watts. These $V_p/\sqrt{2}$ and $I_p/\sqrt{2}$ quantities are called RMS (root mean square) values, and omitting the mathematical theory, leads to the name *root mean square*. The important point is that if we want to make calculations on the power dissipated by a wave, we have to use these RMS

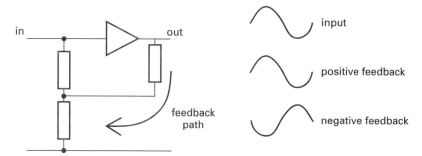

Figure 4.11 Feedback, positive and negative. The feedback signal is a fraction of the output which is added to the input. If this signal is in phase with the input, the feedback is positive; if it is in opposite phase (anti-phase), the feedback is negative.

quantities. For example, if the peak values of a wave are 6 V and 1 A, then the power is not 6 W but only 3 W. This used to be the basis of inflated figures for audio amplifier power outputs, with some manufacturers quoting real RMS power figures and others quoting peak values (not divided by 2) or even rather imaginary values called 'music power'.

All of this assumes that the waves are in phase. If they are not, the amount of true power is calculated by multiplying the RMS voltage by the RMS current and multiplying also by the cosine of the phase angle. If the phase angle is 90°, then the power is zero.

Just in case you thought all this was looking quite logical and orderly, the figure of √2 is true only if the waveshape is that of a sinewave, and different factors (called form factors) have to be used if the waveform is different, as sound waves usually are. This, however, is a worry more for the designers of test equipment than for students of electronics.

For sinewave signals, the true power dissipated in a load can be found by multiplying together the RMS values of voltage and current, assuming that these waves are in phase. The RMS value is equal to the peak value for a sinewave divided by √2, a factor of 1.414. For example, the RMS value of a signal which is 10 V peak is 10/1.414 = 7.07 V.

Feedback

A circuit method that is very important for linear amplifiers is called feedback. Feedback means using a fraction of the output wave of an amplifier and connecting it to the input. This fed-back signal can be connected so that it either adds to the normal input or subtracts from it, Figure 4.11. If the feedback signal is in phase and so is added to the input wave, this is called *positive feedback*, and its action is to increase the gain of the amplifier, reduce its bandwidth, and make the amplifier more sensitive to any changes (such as a small change in the resistance value of a load resistor). If positive feedback is increased to the point where it is enough to provide all the input that the amplifier needs, the result is oscillation – the amplifier will provide a wave output without any input.

Positive feedback is nowadays seldom used in normal amplifiers, but it is the basis of oscillators that are used to generate signals. Until positive feedback was discovered by Armstrong in 1912, radio signals were generated by rotating machines (alternators), and this limited the frequency of signals that could be used. When active devices are used in an oscillating circuit, the range of frequencies that can be generated is limited mainly by the design of the active components, so that it became easy to generate signals at frequencies of 1 MHz and more. Originally, Armstrong's positive feedback was also used to increase the gain of primitive radio receivers, but this type of use was abandoned in the 1930s because of the interference that was caused when users trying to hear a remote station would cause the radio to oscillate and radiate its own signals.

Negative feedback, on the other hand, uses feedback signals that are in opposite phase to the input and it reduces gain. It also increases bandwidth, reduces non-linear behaviour, and makes the amplifier less sensitive to changes in the components. This has made it a standard method that is used by circuit designers who need particularly linear response and wide bandwidth. It is particularly applicable to opamps, see Chapter 3, and it allows the gain of an amplifier to be set by the ratio of two resistors rather than from elaborate calculations with figures that are not particularly reliable.

Signal feedback is extensively used in electronics circuits to modify circuit action. If the gain of an amplifier is more than unity, feeding back a portion of the output in phase to the input will increase gain, reduce linearity and make the amplifier unstable, causing oscillation if the feedback is sufficient. Negative feedback will reduce gain, increase linearity and make the amplifier more stable, provided the phase of the feedback signal remains at 180°.

Oscillators and multipliers

An oscillator is a circuit that is designed to generate signals from a steady voltage supply. Any oscillator can be thought of as an amplifier with positive feedback and some type of circuit to determine the frequency. If this frequency-determining circuit is a resonant circuit, the oscillator is a tuned oscillator and it will generate sinewaves, or at least waves that are close to a sinewave shape. Other circuits can be used that will cause the oscillator to generate a square or pulse waveform. Oscillators can also be designed so that their oscillating frequency can be varied when a steady voltage input is changed, and this type of oscillator is called a voltage controlled oscillator (VCO).

A tuned circuit can also be used as a *frequency multiplier*. Suppose, for example, that an input signal at 1 MHz is applied to an amplifier which is not linear. The effect of the non-linearity is to change the waveshape, and this means that the signal will now contain other frequencies that are multiples of the original. In this example, the output of the amplifier will contain the 1 MHz signals along with others at 2 MHz, 3 MHz, 4 MHz and so on. If the output stage is tuned to 2 MHz, this will become the predominant signal at the output, so that the effect of this stage is of a frequency-multiplier, converting a 1 MHz signal into a 2 MHz signal.

Chapter 5 Block and circuit diagrams

Diagrams

A diagram is a picture that replaces or supplements words as a description of something. One obvious type of diagram is the one that is used for self-assembly furniture, consisting of a set of drawings of the components at each stage of manufacture. At a time when radio technology was the main part of electronics, books and magazines for the amateur used to contain such diagrams, Figure 5.1. This is a *component layout diagram*, showing where the components were fixed and how wires were connected between them – this example is from 1932 and it shows a constructor's diagram for a radio which was, for that time, of advanced design.

The construction of electronics circuits has changed a lot since these days, but diagrams like this are still used, particularly for the home constructor or in servicing manuals so that the precise location of each component can be found quickly. The main difference is that layout diagrams are now

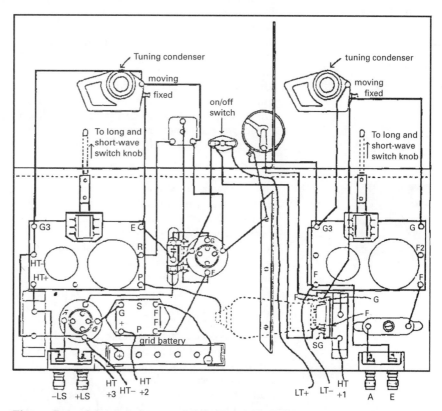

Figure 5.1 A layout diagram of the type that was used by radio constructors in the 1930s. The diagram is useful only if you are using components that correspond to the shapes shown here.

simply an aid to location and are not used as the only information on an electronic circuit.

The trouble with component layout diagrams is that they specify components of a fixed shape. You could not use the diagram of Figure 5.1, for example, if every component had a quite different shape and size, particularly if you did not really know what you were doing. You could not use this type of diagram if you wanted to construct the circuit in a different physical form. You might, for example, want to construct a circuit that had to fit into a space that had been used for a different circuit.

A much more useful type of diagram, both for construction and for design purposes, is one that puts more emphasis on the connections between components rather than on the arrangement of components. This type of diagram is a *circuit diagram*, and Figure 5.2 shows an example taken from a book published in 1973. In this type of diagram, components such as transistors, resistors and capacitors are represented by their standard symbols, and the connections between them are indicated by lines. In this old diagram, lines that are not connected are shown by a loop-over, but more recent diagrams show lines that cross at right angles to mean that no connections are made between these lines. In modern diagrams, all connecting lines are indicated by T-junctions, usually marked by a dot. Components are identified by using letters such as C and R (for capacitor or resistor) along with a number, so that a table of values can be used to find component values. Alternatively, as Figure 5.2 shows, the values could be printed on the diagram itself.

A circuit diagram is not a useful guide to how a circuit will look when the components are put into place and connected up. It does, however, show the connections and how signal and steady currents will flow, so that after a little experience you find such a diagram much more useful than the layout type. For one thing, it can show you how the circuit works; for another, it allows you to construct the circuit in any form you want to use, not simply the fixed pattern that the constructional diagram forces upon you. The circuit diagram is the type that the designer and the constructor will work from. The layout diagram is still useful for servicing, so that you can locate a component which, from looking at the circuit diagram, you suspect might have suffered damage.

Summary

The arrangement of components in a circuit can be shown in a layout diagram, but this is not useful if you are more interested in how the circuit works and what it does. The circuit diagram or *schematic* is a more fundamental type which represents components by symbols and shows connections as lines. A circuit diagram shows the action of the circuit, and need bear no similarity to the layout diagram. The layout diagram is still important for servicing work.

The coming of ICs made circuit diagrams look rather different. There is no point in showing the circuits *inside* the IC, even if you know what they are. ICs are represented in diagrams by squares or rectangles, with connections numbered or lettered. Figure 5.3 shows a typical diagram of the 1980s, taken from the standard textbook *Servicing Electronic Systems*. This diagram shows the IC as a rectangle with 24 connecting pins, and ordinary (*discrete*) components connected to these pins. The circuits inside the IC, if we could show them, would probably require several pages of close-packed diagrams. This takes us one step backwards in understanding, because you

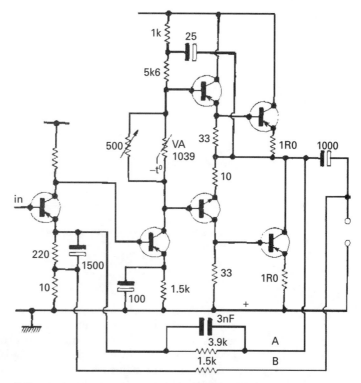

Figure 5.2 A circuit diagram of the 1970s, showing the connections of the circuit. Components are represented by symbols and connections by joining lines. This form of diagram allows you to use whatever physical layout you prefer.

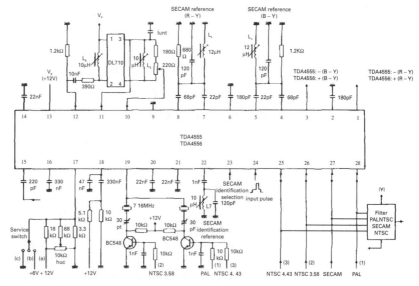

Figure 5.3 A typical modern diagram in which the main component is an IC. The circuit paths are all inputs or outputs.

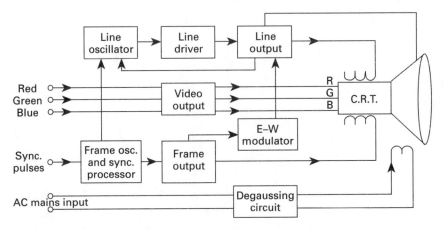

Figure 5.4 A modern block diagram. This example is of a monitor for a computer.

can no longer work out how the circuit behaves simply by looking at the circuit diagram; you need also to know what the IC does, in terms of what inputs it requires and what outputs it provides. A modern circuit diagram may use only one or two ICs, so that a layout diagram is superfluous, but it must be accompanied by some information on the ICs.

Circuit diagrams are useful for designers and for servicing work, but for anyone without a specialised knowledge of electronics they do not contain the right type of information. In particular, looking at a circuit diagram, even of the older type, does not tell you about what a circuit achieves unless you already know a considerable amount about electronic circuits. There is just too much detail on this type of diagram, and the old adage about not seeing the wood for the trees applies. What you need to understand electronics methods is a diagram that concentrates on ends rather than on means, on signals, rather than the arrangement of components. Such a diagram is called a *block diagram*.

Block diagrams

To start with, the shapes that you see in a block diagram are, not surprisingly (but not always), blocks. Each block shows signals in and signals out, along with a name for the block and, very often, small sketches of the signal waveforms. All power supplies are ignored, as are individual components. At one time, each block would have corresponded to a single IC, but for some block diagrams this would not be enough and several blocks might be needed to explain the action of a single IC – after all, the whole of your circuit might be contained in one single IC nowadays. Figure 5.4 shows a typical modern block diagram.

Now you might think that Figure 5.4 is not a great advance. It is a block diagram for a computer monitor, and it brings home the point that no picture can completely replace words. If you are to make sense of a block diagram, you must know what the circuit is intended for and how it deals with the actions. The block diagram for a computer monitor does not mean a lot until you know what input signals exist and how they are used to affect a cathode-ray tube. We'll look at these aspects later, but the point is important – words and pictures go together and both are needed; you cannot learn from block diagrams alone.

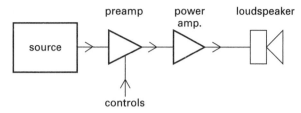

Figure 5.5 A simple block diagram for an audio circuit.

Electronic devices can be drawn using diagrams. A component layout diagram is useful for constructional and servicing work, a circuit diagram is used for design and trouble-shooting, and a block diagram is used to understand quickly how a device works. The block diagram uses few symbols and shows the effect of a circuit by sketching the input and output waveforms.

Linear circuit blocks

Figure 5.5 shows a simple linear circuit block diagram. In this example, the triangular shapes are used to mean linear amplifier circuits as an alternative to the rectangular block shape. The diagram is intended to show a simple audio amplifier that takes an input signal from a source (tape, CD, radio), amplifies the wave amplitude (voltage), applies volume and other controls, and then uses a power amplifier to produce an output to a loudspeaker.

This illustrates the importance of words along with the diagram. Amplification, remember, really means making a copy of a waveform, and the copy usually is scaled up, with a greater amplitude of voltage. A circuit that carries out this action is a voltage amplifier, and an opamp is one example. Voltage amplification is always needed because the amplitude of signals from most sources like microphones, tape players, CD players and so on is very small, anything from a few microvolts to a few millivolts. Once a wave has been amplified to a volt or so, we can use controls like volume controls, bass and treble controls and so on. This part of an amplifier system is often called a *preamplifier*.

Why not use these controls on the smaller amplitude signals? The reason is *noise*. A signal is caused by the movement of electrons, but electrons are always in some kind of movement, making random shuffling motions in any conducting material – the hotter the material, the more the electrons vibrate (and that's what we mean by temperature). If we connected an amplifier to any piece of conducting material and increased the amount of amplification (gain), the output would be a sort of signal, a jagged shape with no repeating pattern that we call electrical noise. When this signal is applied to a loudspeaker, it sounds like a rushing noise, and when it is applied to a TV screen, it looks like white speckles on a black sheet.

Volume controls make rather more noise than other passive components, and we always try to avoid placing volume controls right at the start of an amplifier, because then the noise that they create will be amplified. If we amplify up the feeble signal that we want, and *then* use the volume control, we can ensure that the noise of the volume control is much less than the (wanted) signal amplitude. In electronics talk, we have a large *signal-to-noise ratio* (written as S/N). Where a large amount of linear amplification is needed, we use a voltage amplifier (a preamplifier) first, one that is designed to add as little noise as possible. We can then use our volume and other controls on this amplified signal.

Linear ICs have steadily developed over the years, and it would be

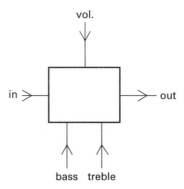

Figure 5.6 A way of representing control inputs in a block diagram.

Table 5.1 Gain ratios and decibels

Gain	dB	Gain	dB	Gain	dB	Gain	dB
2	6	20	26	200	46	2 000	66
3	9.5	30	30	300	50	3 000	70
5	14	50	34	500	54	5 000	74
10	20	100	40	1 000	60	10 000	80

Note: values have been rounded. You can use this table to find intermediate value by remembering that *multiplication* of gains equals *addition* of decibels. For example, a gain of 15 times is 5 × 3, which in dB is 14 + 9.5 = 23.5

unusual to find separate preamplifier and power ICs nowadays. Accordingly, it's less usual to see the triangular symbol used on a block diagram because of the number of inputs and outputs. The main input, of course, is the feeble signal from the tapehead, pick-up, microphone or whatever, but there are also connections for volume and other controls. At one time, these would be signal outputs and inputs that could be connected to a variable potentiometer, but on modern circuits the control is carried out using a transistor circuit which uses a DC input. In other words, the higher the steady voltage applied to the circuit, the more signal it passes. This system ensures that the signal is processed entirely inside the IC, so that there is less risk of interfering signals being picked up at the control terminals, or of noise from the controls. This does not mean that the noise problem is any less, because there will inevitably be some noise from the internal circuits that carry out the volume control action. Active components are always likely to generate more noise than passive components, but the noise is easier to deal with than the noise from a volume control.

The point about block diagrams, however, is that they need not show such detail. If you want to show a block for a circuit that carries out voltage amplification, you simply draw a rectangle, show input and output points, and indicate what it does, using lines to show that there is provision for volume or other controls, Figure 5.6. More information is usually needed, and for a voltage amplifier, for example, you will need to know what amount of amplification (or *voltage gain*) is being achieved.

An amplifier block, then, is very likely to be labelled with its figure of gain expressed in decibels. A gain of 40 dB corresponds to a voltage gain of 100 (and a power gain of 10 000). Table 5.1 shows some values of voltage

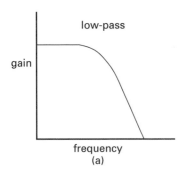

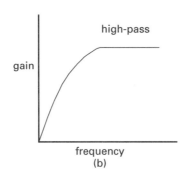

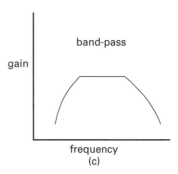

Figure 5.7 Gain–frequency or *characteristic* graphs for three basic types of filters.

gain ratios and decibel amounts. The advantage of using the decibel scale, apart from the fact that the numbers are smaller, is that it corresponds better to human senses. One decibel, for example, is pretty close to the smallest difference that your senses can detect.

Block diagrams omit all the detail of a circuit diagram and bear no relation to component layout. They are concerned with the actions that are performed on signals, and the usual system is to use one block for each action. Where this would lead to a diagram being too large, a set of actions can be represented by a block. The block diagram must show the inputs and outputs for each block, and this is often supplemented by other information, such as a name, and by waveform or timing diagrams.

Filtering

There is one important action that is very commonly used in linear circuits and which often has to be represented on block diagrams. *Filtering* means selecting one range of frequencies from others when there is a mixture of different signals present. You might, for example, want to filter a mixed signal so that it contained only the frequencies that your ear would detect if the signal was used to operate a loudspeaker or headphones. You might want to filter a mixed signal so that it contained only the higher frequencies, free from power-line frequencies. You might also want to filter a mixed signal so that all the higher frequencies are removed.

A filter is an electronic circuit that acts selectively on one frequency or a range of frequencies, either to pass or to reject that frequency or range.

When a filter passes only the lower range of the frequencies that are applied to it, we call it a *low-pass filter*. A typical graph of output plotted against input for such a filter is illustrated in Figure 5.7(a). You can also illustrate a graph for the response of a *high-pass filter* (Figure 5.b) which rejects the lower frequencies of a mixture and passes only the higher frequencies. The other basic type of filter is the *band-pass type* (Figure 5.c) which rejects both the highest and the lowest frequencies and passes only a range between these extremes. A radio tuner carries out this type of band-pass action so as to receive only one station and not a garbled mixture of all.

high-pass   low-pass   band-pass   band-stop

Figure 5.8 The symbols that are used to mark a filter in a block diagram.

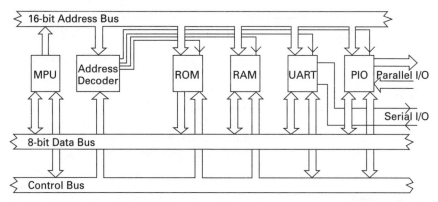

Figure 5.9 A computer block diagram, showing the representation of *buses*. This diagram will appear again in Chapter 12.

You can also use a band-stop filter to reject a range of frequencies, and another possibility is a spot filter (or sharp-cut filter) which rejects one particular frequency along with a small range of frequencies around it.

This sort of illustration is too much for a block diagram (though you might want to include it as part of the text explaining the block diagram). On a block diagram, we use a set of wave symbols to illustrate what a filter does. The single wave means low frequencies, the double wave represents middle frequencies, and the triple wave represents high frequencies. By placing a diagonal bar across a wave symbol we mean that the filter rejects this range of frequencies. Figure 5.8 shows some filter symbols used to indicate the basic types of filters.

Filters are important circuit blocks that appear when a frequency or range of frequencies must be singled out for use. The most common filter types are low-pass, high-pass and band-pass, but band-stop and sharp-cut filters are also used.

Digital circuit blocks

Digital circuit blocks make use of the same type of rectangular shapes as can be used for the more elaborate linear circuits. The action of the circuits is likely to be very different; for example, we are very seldom interested in amplification, but nearly always concerned with timing. The symbols that we show along with the block shapes are therefore very different.

Take a look first at a block diagram, Figure 5.9, for a computer, the standard diagram which is used in City & Guilds examinations for electronic servicing courses. Each block is of the same size on this diagram, and the names indicate the main parts of the computer (this is the main computer,

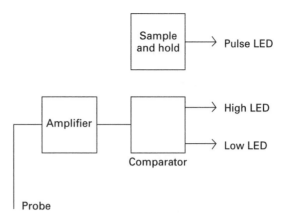

Figure 5.10   A block diagram for a logic probe (used for checking digital circuits).

not including items like keyboard, printer, monitor, etc.). The main difference between this and linear circuit block diagrams (or between this and many other types of digital circuit block diagrams) is that there is no clear and obvious input and output. We'll look at this point (and this block) again in Chapter 12, but first of all, note the use of arrows.

The arrows show the direction of signals, and where you see double-headed arrows, this means that signals will be passing in opposite directions, usually at different times or on different lines. The fact that these are broad (or block) arrows is also significant. When we want to indicate a single signal line, we can use a thin line with an arrowhead – there are some in this example. The outline arrows are used to indicate a set of signals. For example, we might have a set of 32 signals used as a signal connection to a block. Rather than show the individual lines (which goes against the principle of using a block diagram to eliminate detail) we simply draw a 'fat arrow'. When these lines are being used at one instant to provide input signals and at another instant to provide outputs, we indicate this by using the double arrowheads, one at each end of the thick arrow.

Even if you know absolutely nothing about a computer, you can see from this diagram that, for example, the block marked MPU will send signals to the part called 'Address bus', and will both send to and receive from the part called 'Data bus'. A *bus* is, in fact, a collection of lines that can pass signals in either direction, so that when any block sends signals to a bus, anything else connected to that bus can make use of the signals. This is a type of connection that is used to a considerable extent in digital circuits to make connections simpler.

Now we consider something which is lower down in the complication scale. Figure 5.10 shows a block diagram for a logic probe, an instrument that is used for trouble-shooting digital circuits. The principle is to connect a probe-lead to a line in a digital circuit so that the logic probe can indicate whether that line is at a low voltage (usually zero volts), at a high voltage (meaning +5 V or +12 V), or carrying pulses.

In this diagram, the connection is indicated by the line labelled *probe*. The signal on this line is the input to a block marked *amplifier*, but this is not a voltage amplifier in the sense that we would use in a linear circuit. This amplifier is an example of the type of circuit that is called a *buffer*. It ensures that the power that is needed by the instrument is not taken from the circuit that is being tested. The voltage at the test point is the input to this

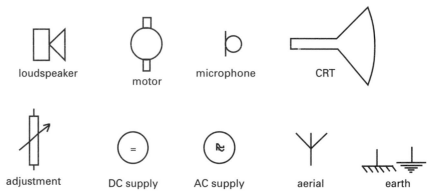

Figure 5.11 Some symbols that are used along with block representations.

amplifier, and the voltage out is the same. Whatever power is needed at the output is supplied by the amplifier, not by the circuit that is being tested.

The next block is labelled *comparator*. This compares the voltage from the amplifier (buffer) to find if this is high or low (the meaning of high and low will be clearer after you have read Chapter 10). If the sampled voltage is high, one of the outputs from the comparator will activate a light, the High LED. If the output is low, the other main output is used to activate the Low LED light. If the output is pulsing, changing between high and low, this output is taken to the block marked *sample and hold*, and used to light a third LED.

Block diagrams for digital circuits use the same block shapes, but the connections are usually multiple lines called buses. The precise waveforms are not important, but timing is, and very often it is impossible to show the details of timing on the block diagram.

Other blocks

Everything in a block diagram can be represented by blocks, but it is often more convenient to use a few circuit symbols that can be recognised. For example, if a linear circuit drives a loudspeaker, the block diagram can show the loudspeaker symbol because this is easy to recognise, and your eye will pick it up better than a box marked L/S (for example). Other symbols that are often used are for a cathode-ray tube (CRT), a motor or a potentiometer, and some of these are illustrated in Figure 5.11.

Chapter 6   How radio works

Radio waves

We use the phrase *radio waves* to mean electromagnetic waves that are transmitted across space (not just through air) as distinct from waves along wires. The existence of radio waves was predicted by Clerk Maxwell and later discovered experimentally by Heinrich Hertz, but no use was made of the waves until Marconi (along with Popov and others) used them for communications. They have come a long way in a time of about a hundred years.

We now know fairly well what happens. When we cause a voltage to exist between any two points in space, the *space* itself is distorted – we refer to the effect as an *electric field*. A varying voltage causes a variable distortion which appears to us as magnetism – we say that a *magnetic field* exists, meaning that this piece of space causes magnetic effects. This is a changing magnetic field, however, and its effect is to generate another voltage some distance away, which in turn causes a magnetic wave and yet another voltage wave. These waves are some distance apart, because they travel away from the point where they were generated at a speed of around 300 000 000 metres per second, which is also the speed that has been measured for light. No air or any other material is needed to transmit these electromagnetic waves, though they can pass through insulating materials and they can also be conducted along metals and other conducting materials. Electromagnetic waves can be reflected and refracted and they can also be polarised (see later), all effects that are familiar from our knowledge of how light behaves.

In the early days of radio, experts believed that long-distance communication was impossible because of the curvature of the earth. They would have known better if they had followed the work of Oliver Heaviside, who worked out that the effect of the radiation from the sun would be to split atoms of gas in the upper atmosphere (the *ionosphere*) and provide a conducting layer that would reflect waves. Heaviside, however, was regarded as an eccentric engineer and was scorned by academics, and much of his work was not published until after his death. Marconi and others disregarded the academics and pressed on with radio contacts over greater and greater distances, culminating in the transatlantic transmission in 1901. Academics then had to rediscover all that Heaviside had done. We'll come back to the effects of the ionosphere later.

All electromagnetic waves travel at the same speed in space, and the speed, around 300 000 000 metres per second (3×10^8 m/s), is almost the same in air. In addition, these waves also have a measurable *wavelength*, meaning the distance from the peak of one wave to the peak of the next. As for water waves and sound waves, the speed, frequency and wavelength are related by the equations:

$$v = f\lambda \text{ or } f = v/\lambda \text{ or } \lambda = v/f$$

with f for frequency in Hertz, v for speed in metres per second and λ wavelength in metres. For example, if you generate a signal at 1 MHz, then

Table 6.1 Wavelength and frequency

Frequency	Wavelength	Frequency	Wavelength
100 kHz	3 km	50 MHz	6 m
1 MHz	300 m	100 MHz	3 m
5 MHz	60 m	400 MHz	75 cm
10 MHz	30 m	1 GHz	30 cm
20 MHz	15 m	10 GHz	3 cm

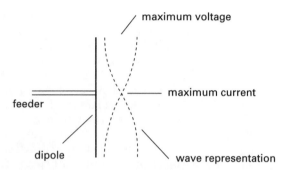

Figure 6.1 Dipole aerial principles. The dashed lines represent the half-wave of signal at the aerial, ensuring the maximum signal current to the feeder line of a receiver (or from the feeder of a transmitter).

its wavelength in space is 300 000 000 divided by 1 000 000, which is 300 metres. A wave at 100 MHz, such as we use for FM radio (particularly in the Torquay area), has a wavelength of 300 000 000/100 000 000 which is 3 metres. Table 6.1 shows a few examples of frequencies and wavelengths.

Radio waves are electromagnetic waves in space (which includes air) travelling at around 300 million metres per second. The frequency of a wave is the number of oscillations per second, and this causes a wave to have a wavelength which is found from speed/frequency.

It's not difficult to generate and launch waves in the frequency range that we use for radio. Anything that produces an alternating voltage will generate the signals, and any piece of wire will allow them to radiate into space. We find, however, that a launching wire, which we call an *aerial*, works best when its length is an even fraction of the wavelength of the wave. Lengths of half a wavelength or quarter of a wavelength are particularly useful, and a very efficient system, called a dipole, has been used for many years. These tuned aerials are not so efficient (and are difficult to construct) for the very long wavelengths, and they really come into their own for wavelengths of a few metres or less. Though we still make considerable use of the very long wavelengths, most of the communications equipment (FM radio, television, mobile phones) that we use nowadays depends on using the short wavelengths.

A dipole aerial, Figure 6.1, consists of two sections, each one quarter of a wavelength long. If you imagine the inside portions connected to a signal, the voltage on the aerial will follow a wave pattern, and when the voltage is almost zero at the centre it will be a maximum at the ends. This is the type of aerial that is very familiar to us and is used for TV and FM radio reception, with the rods cut to the correct quarter-wavelength size for the nearest set of transmitters. The dipole is most efficient for the wave that is exactly of the size, but this isn't too critical, particularly if the rods of the dipole are thick. The points where the feeder cable is attached correspond to maximum signal current, and should be connected to a low resistance.

Dipoles are also polarised. Polarisation, as far as radio aerials is concerned, means the direction of the electric field of the wave, and this is often fixed (by the design of the transmitting aerial) as vertical or horizontal. If the transmitter uses a vertical dipole, the signal will be vertically polarised, and the receiver aerial should also be vertically polarised. If the transmitter uses horizontal polarisation, the receiver aerial should also be horizontally polarised. Different polarisation directions are used to reduce interference between transmitters that use the same frequency, and in some cases, because one direction of polarisation provided better reception conditions in difficult terrain than the other.

Where the length of a dipole might be excessive, or if a dipole would be inconvenient (as for a car phone, for example), we can use the upper half of a dipole, often referred to as a *whip aerial.*

Radio waves all travel at the same speed in space, and have wavelength and frequency. These quantities are related, and speed = wavelength × frequency. A radio wave is most efficiently radiated from a metal aerial with a length that is a suitable fraction of a wavelength, such as half-wave, and reception of radio waves is best when the receiving aerial is also a suitable fraction of a wavelength. For low frequencies with wavelengths of several kilometres, half-wave aerials are impossible, and it's fortunate that half-wave aerials are most needed for waves where the wavelengths are less than a couple of metres. Dipole aerials are polarised and their direction of polarisation should match that of the transmitter.

Transmission and reception

Early radio transmitters used sparks to generate signals (which were a mixture of frequencies, mainly in the UHF range), and at a short range, these were seen to cause sparks to appear across the two sections of a receiving aerial. More sensitive ways of detecting the received signal soon followed, and a device called a coherer, consisting of metal filings in a glass tube, was used extensively, along with headphones, in the early days of radio as a way of detecting signals that were too feeble to make a spark. For some time, large land-based transmitters used mechanical generators, alternators, working at about 20 kHz, and with huge aerials so that their signals could span the Atlantic. These and spark generators were used until radio valves were invented around 1906, though spark transmitters were still in use when the *Titanic* sank in 1912, and her distress signals were picked up in New York by a Marconi radio operator called David Sarnoff (who later founded the Radio Corporation of America when the US government threatened Marconi with the equivalent of nationalisation).

Reception of radio waves is simple enough – make a dipole or a single-wire aerial, and detect the small currents at the centre terminals. Detection is a snag, though. At the start of the twentieth century, there was nothing that could amplify a signal at 20 kHz, and devices like earphones cannot respond to such higher frequencies or the much higher frequencies that were generated from sparks. The only way that the presence of the wave

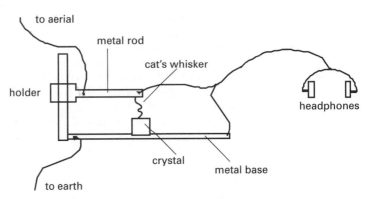

Figure 6.2 A crystal set. This vintage radio diagram shows the use of a crystal and a fine wire (cat's whisker) to demodulate radio waves into the audio signal.

could be detected was by what we now call *rectification*, using some device that would allow the current to flow only one way. The coherer with its metal filings was the first attempt to make such a device, but later it was discovered that crystals of the mineral Galena could be used. The crystal was placed in a metal holder connected to earphones, and an aerial was connected to a fine wire, called a cat's whisker, jabbed against the crystal, Figure 6.2.

When a radio wave hit the aerial, the alternating current flowed only one way through the coherer or the crystal, and the headphones gave a feeble click, which sounded reasonably loud if you happened to be wearing the headphones. If the radio waves had been switched on and off again quickly, the sound in the headphones was a short double-click; if the wave was sustained for longer the gap between clicks was longer. By switching the transmitter on and off using a Morse key, a skilled listener could read the short (dot) and long (dash) intervals between clicks and write down the Morse code message that was being transmitted. This was the basis of radio as we used it right into the 1920s.

Anything that can generate oscillations can form the basis of a transmitter, and early designs used sparks, generating a huge mixture of wavelengths. Later designs used mechanical generators for low-frequency waves that required huge aerials. Communication was carried out by using Morse code, switching the signals on and off in the familiar dot and dash patterns.

Neither coherer nor crystal was really satisfactory. The coherer had to be tapped regularly to settle the filings into place, and the cat's whisker had to be moved over the crystal surface to find a 'sweet spot', a place where the action was most effective. This movement of the cat's whisker often had to be repeated, usually just as you particularly wanted to get the important part of a message.

Modulation

Morse code served radio well and is still useful for emergency purposes because it requires the absolute minimum of equipment and it is the most efficient use of radio – even a low-power transmitter can have a long range

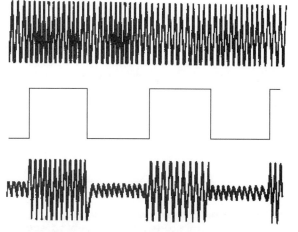

unmodulated carrier – a high frequency sinewave with a constant amplitude

square wave modulation

modulated carrier – the amplitude of the sinewave rises and falls in the pattern of the square wave and at the same frequency

Figure 6.3 Amplitude modulation. The signal, shown here as a square wave, alters the shape of the carrier so that the resulting outline (the *envelope*) has the same shape as the modulating wave.

using Morse. It's not exactly useful for entertainment purposes, however, or for long-distance telephone calls, and though a method of carrying sound signals over radio waves had been demonstrated early on (on Christmas Day in 1906), it took some time to make this a commercial proposition (in 1920, in fact). The use of radio waves to carry the information of sound waves requires what we call *modulation*, and in the early days of radio this was always the type we call *amplitude modulation*.

To start with, we have to change the sounds, whether talking or music, into electrical waves, using a type of transducer called a *microphone*. Microphones were, thanks to the use of telephones, reasonably familiar devices in 1906 (in the USA at least), so that the only obstacle to broadcasting in these early days was knowing how to make a radio wave of 100 kHz or more carry a wave, the audio signal, that is of a much lower frequency, about 300 Hz. The simplest answer is to make the amplitude of the radio wave rise and fall in sympathy with the lower frequency, Figure 6.3. This is what we mean by *amplitude modulation* – the amplitude of the carrier wave is controlled by the audio signal.

Amplitude modulation had been possible even in the early days of spark transmitters, but only with difficulty and using primitive equipment. The use of radio valves, called vacuum tubes, or just 'tubes', in the USA, made efficient modulation possible, because the valve allows a current to be controlled by a small voltage. In addition, another valve can be used as an oscillator, generating the radio waves and replacing the mechanical or spark type of generators. The first radio station intended for entertainment started operation in November 1920 at Pittsburgh, Pennsylvania. The usual way of modulating was to control the DC supply to the output valve of the transmitter, using another valve, and making the input of this valve the audio signal.

When a radio wave is modulated in this way, it is no longer a simple wave. A graph of the waveform would show the type of shape illustrated in Figure 6.3, but there is another effect that is shown only when the signal amplitude is compared with frequency rather than time.

Suppose, for example, that we modulate 100 kHz radio waves with a 10 kHz sinewave signal. Analysing the frequencies that are present shows

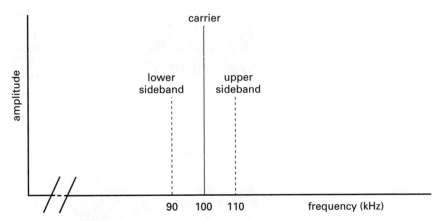

Figure 6.4 Sidebands. A carrier can be represented as a vertical line on a graph of amplitude plotted against frequency. When this carrier is modulated by a sinewave, two more frequencies appear, the upper and lower sidebands. The transmitter must provide the extra power represented by these additional frequencies.

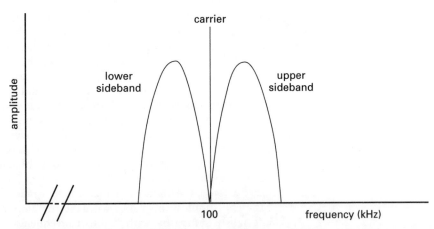

Figure 6.5 Audio signal sidebands. When the carrier is modulated by a range of frequencies, as it would be for an audio signal, the sidebands cover a range of frequencies which is equal to the audio bandwidth. Both upper and lower sidebands contain the same signal information.

that three main waves now exist. One is at 100 kHz and is the carrier wave. There is also a wave with a frequency of 90 kHz, called the *lower sideband*, and one with a frequency of 110 kHz, called the *upper sideband*. These sidebands appear only when the carrier is being modulated, and their amplitude is less than that of the carrier, Figure 6.4. In general, if the carrier frequency is F and the signal frequency is f, the sidebands are at $F - f$ and $F + f$, so that the total bandwidth of the transmitted wave is from $F - f$ to $F + f$, a bandwidth of $2f$.

If we modulate the carrier with a signal that consists of a mixture of frequencies up to 10 kHz in bandwidth, the sidebands will show a pattern such as appears in Figure 6.5, showing a spread of sideband frequencies rather than two single frequencies. In each sideband, the frequencies closest to the carrier frequencies are due to the lower frequencies of the modulating signal, and the frequencies furthest from the carrier frequency are due to the higher frequencies of the modulating signal.

All we need to carry the information is one of these sidebands, so that both the carrier frequency and the other sideband are redundant. There are ways that can be used of eliminating the carrier (*suppressed carrier* transmission) and one sideband (*single-sideband* transmission) from the transmitted signal, but for entertainment purposes the additional complications at the receiver make these methods undesirable, though both stereo radio and colour TV make some use of these methods for additional signals (using sub-carriers, see later).

Summary

Modulating a radio carrier wave always causes sidebands to appear. For amplitude modulation, the difference between a sideband frequency and the carrier frequency is equal to the modulating frequency. A modulated wave therefore requires a greater bandwidth than an unmodulated wave, and its efficiency is low, because the carrier and one sideband contribute nothing – all of the information is contained in a sideband (either sideband).

Reception

At the receiving end, valves were still a far-off dream for the few people in 1920, mainly amateur enthusiasts, who could get hold of crystals and headphones and rig up an aerial. Some 1000 listeners probably heard the 1920 US presidential election results in this way, and in the years to come this number multiplied rapidly. Even in controlled and restricted Britain, listeners experimented, trying to hear experimental transmissions from the Marconi workshop in Writtle (Essex) and from continental broadcasts when conditions and geographical position suited. In 1923, a private company, the British Broadcasting Company, was formed and used nine transmitters.

By the 1930s, radio as we know it was a reality. The BBC had been nationalised so that the government could control it and tax it, but in the USA it remained free in every sense and consequently the US remained at the front of technical (though not necessarily artistic) progress. Small valves became available and were used in receivers, and by the 1930s loudspeakers were normally used in place of headphones so that the whole family could listen in. Amplitude modulation was used for transmissions with frequencies in the medium-wave band of about 530 kHz to 1.3 MHz, and the receivers developed from the simple one-valve pattern to multi-valve types, some of the expensive models featuring an added electronic record-player added to make a radiogram. By the late 1930s a few lucky people in the London area could, if they could afford it, buy a combined radio and TV receiver (though TV was confined to a couple of hours in the evening, and a receiver cost as much as a small bungalow).

A severe problem with the use of AM and the medium waveband was interference between stations. Medium-wave broadcasts can be picked up over quite long distances, up to 1000 miles, and where two stations used frequencies that were close, they set up an interference that could be heard as a whistle at a receiver tuned to either station. The whistle, in fact, is at the frequency that is the difference between the two transmitted frequencies. For example, if one station transmits at 600 kHz and another at 605 kHz, the whistle is heard at a frequency of 5 kHz. As radio transmitters were built all over the world, it became quite impossible to share out frequencies on the medium-wave band, and interference became the common experience for all listeners, particularly after dark when the reflecting layers in the ionosphere moved higher and reflected waves that originated from greater distances.

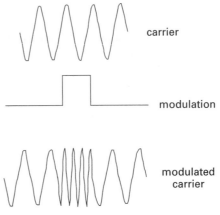

carrier

modulation

modulated
carrier

Figure 6.6 Principle of frequency modulation. The amplitude of the carrier remains constant, and the signal alters the frequency of the carrier. The amount of alteration is called the deviation of the carrier. The bandwidth of sidebands created by this type of modulation is larger than that created by amplitude modulation.

A way of getting around these problems had been devised by an inventor who, of all that ill-treated fraternity, was the least fortunate. Edwin Armstrong developed *frequency modulation* (FM) to eliminate the effects of interference, both man-made and natural, and built a radio station (now a working museum) in the Catskill mountains just outside New York to prove the point. As the name suggests, the low-frequency audio signal is carried by altering the *frequency* of the radio signal, Figure 6.6, with the amplitude kept constant. Since interference of all types only alters the amplitude of radio waves, FM can be virtually free from interference unless it is so severe as to reduce the radio wave amplitude to almost zero. FM is used universally today for high-quality broadcasts and for short-range local radio because if two FM transmitters are using almost the same frequency, a receiver will lock on to the stronger transmission, ignoring the other one and with no whistle effects.

Medium-wave broadcasting became so popular that by the later 1930s there were too many transmitters to permit a reasonable bandwidth to be used by each, and the interference between transmitters that were on adjacent frequencies caused whistling noises in receivers. The solution was to use a different modulation system, FM, and to transmit FM signals using high-frequency carriers, typically in the 100 MHz region.

Summary

Radio blocks

Early receivers

Early radio receivers for AM signals consisted of an aerial, a crystal and headphones. This could be used close to a transmitter, and when a coil and capacitor was added it became possible to tune to more than one transmitter signal. The crystal is the detector or demodulator that allows the low-frequency signal to be extracted from the radio wave, and the principles now are the same as they were then. When a modulated signal is passed through any device that allows only one-way traffic (a *demodulator*), only half of the modulated wave will pass through. As Figure 6.7 shows, this makes the signal asymmetrical, and a small-value capacitor between this point and earth (together with the resistance of the earphones) will integrate the signal, ignoring the rise and fall of the radio frequency waves and following only the modulation signal.

principles:

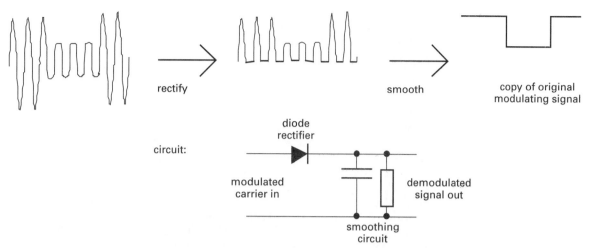

circuit:

Figure 6.7 Principles of amplitude demodulation (detection). The rectifier passes only one half of the modulated carrier, and filtering leaves only the outline which corresponds to the modulating signal.

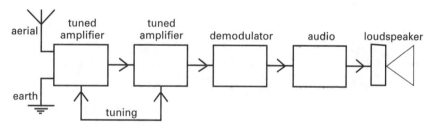

Figure 6.8 A block diagram for a TRF receiver, the first type of receiver using valves.

All radio receivers have developed from this simple beginning, and the two aims of development have been to improve both *sensitivity* and *selectivity*. Sensitivity means the ability to pick up and use faint signals from remote or low-power transmitters. Selectivity means the ability to separate radio signals that are of adjacent frequencies. Both are important if you want to use a radio with a large choice of transmissions.

Sensitivity requires amplification, and selectivity requires tuning. Though this was well understood in the early days of radio, there were always problems. For example, if you amplify a radio wave too much there is a danger of *positive feedback*, causing the receiver to oscillate, and drowning out reception for all other receivers near it. If a receiver is too selective, the sound that you hear is unintelligible, lacking the higher frequencies. In the phrase that was used at the time, it is a 'mellow bellow'.

By the start of the 1930s, a typical radio receiver followed a design called the TRF, meaning 'tuned radio frequency', for which the block diagram is shown in Figure 6.8. The feeble signal from the aerial is both amplified and tuned in one or more amplifying stages that used valves. The tuning circuit used a coil and a variable capacitor (two sets of metal vanes separated by air and arranged so that they could mesh in and out). If more than one tuned circuit was used for greater selectivity, the tuning capacitors had to be 'ganged', meaning that they could be moved in step with each other, using a single metal shaft to carry all the moving vanes of all the capacitors, and a single control. The sensitivity of these radios was often improved by a small amount of positive feedback of the radio signal, and the crystal that had been used in the early days was replaced by another valve, a *diode*, that provided demodulation.

The action of the diode demodulator is the same as that of the older crystal, but with the advantage that valve diodes could be mass-produced and were reliable. The output signal from the diode was still feeble, more suited to earphones than to a loudspeaker, so that the natural path of development was to add another valve amplifier for the low-frequency audio signal, usually with a low-pass filter to get rid of the remains of the radio-frequency signal and so prevent it from being fed back to earlier stages.

These radios were a very considerable improvement, both in sensitivity and selectivity, on the old crystal sets, particularly when more than one stage of tuning was used, but as the medium waveband started to become crowded the old problems returned. Selectivity was still not enough, and attempts to increase sensitivity caused positive feedback and 'howling'. This latter problem was looked on as highly anti-social, and some remedy had to be found. Attempts to make radios with three or more tuned circuits made the problem worse, because with three tuning capacitors on one shaft, there was always a path for positive feedback of waves from the output to the input.

The most primitive radio system following the crystal set was the TRF receiver. A set of tuned circuits along with amplifier stages were used to select and amplify the wanted frequency. This amplified signal was then demodulated and the audio signal further amplified to drive a loudspeaker. The disadvantage was that the tuned circuits had to be ganged, which made it impossible to isolate them from each other, so that positive feedback and oscillation were always a problem.

The superhet

The solution to the problems of medium-wave radio lay in an earlier invention by Edwin Armstrong. This bore the full title of the supersonic heterodyne receiver, mercifully abbreviated to '*superhet*', and it is still the type of circuit we use for all reception of radio, TV and radar today. The principle is to eliminate as far as possible the amplification of the carrier frequency, so that variable tuning is used only at the start of the receiver block. The invention deliberately makes ingenious use of the 'whistle' frequency that is generated when two radio frequencies are mixed together.

The principle, as it is used in medium-wave radios, is illustrated in Figure 6.9. A tuned circuit selects a range of carrier frequencies, and this is used as the input to an amplifier. At the same time, another frequency is generated in an *oscillator* circuit and applied to the same amplifier. This generated frequency is not the same as the received frequency, it is exactly 550 kHz higher, and the tuning capacitor for the oscillator is ganged to the input tuning capacitor so that these frequencies stay exactly 550 kHz apart.

The block diagram shows actions, and these were often combined in older radios, so that, for example, the tuned amplifier, oscillator and mixer actions were often carried out by one radio valve. Later, actions were separated and each action was carried out by a separate transistor.

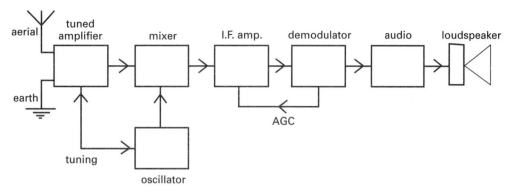

Figure 6.9 A block diagram for a typical superhet receiver, showing the AGC connection.

The result is that the output of the first stage, called the *mixer*, contains four lots of radio signals. Suppose, for example, that the incoming radio wave is at 700 kHz and is amplitude modulated. The oscillator will generate an unmodulated 1250 kHz radio wave, and the result is that the output of the amplifier consists of waves of 700 kHz, 1250 kHz, 550 kHz and 1950 kHz. These are the input signals plus the sum and difference of the frequencies. These sum and difference frequencies are modulated like the input frequency.

Now the 550 kHz signal is easy to separate by a filter, and it can be amplified. Any feedback of this signal to the input of the amplifier is not likely to cause much harm, because it is at a very different frequency from either the input wave or the generated wave. In addition, because this new frequency, called the *intermediate frequency* (IF) is fixed, it can be tuned by circuits that are fixed; there is no need to try to alter the tuning of these circuits when you tune from one station to another. In addition, metal boxes can be put over the IF tuned circuits to reduce any feedback even further. Adding more IF stages dramatically increases both selectivity (because there are more tuned circuits) and sensitivity (because there are more amplifier stages), so solving, for quite a long time, the problems of the crowded medium waves.

One feature that was used more and more even in the early days is *automatic gain control* (AGC). The superhet is a sensitive receiver, and if it is sensitive enough to provide a useable output from the faint faraway transmitters then the nearby ones are likely to overload it, causing severe distortion. In addition, because radio waves are reflected from shifting layers of charged particles in the sky (the ionosphere), the received signal usually fluctuates in strength unless it comes from a nearby transmitter. Figure 6.10 illustrates this, showing a wave that can reach a receiver by two paths, one of which is a reflected path. At any instant, these two waves can be in or out of phase. When they are perfectly in phase, they add so that the signal strength is increased compared to the strength of a single wave. When the waves are out of phase, the signal strength is reduced. Because the reflecting layers in the atmosphere are charged particles located a few hundred miles above the surface of the earth and continually moving, the phase of the reflected signal is constantly changing, and so the received signal strength continually fluctuates.

This is less of a problem for FM radio and for TV signals, because the higher frequencies that these services use are not reflected to anything like the same extent by the layers in the upper atmosphere, and only the direct wave is used over a comparatively short range (100 miles or less). Occasionally,

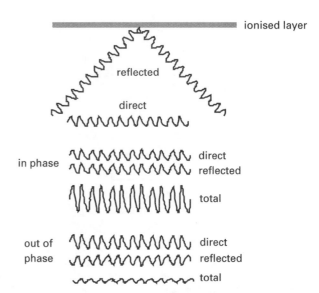

Figure 6.10 Why AGC is needed. Unless a receiver is very close to the transmitter, a mixture of direct and reflected waves will be picked up, and the phase of the reflected waves will vary as the charged layers (ionosphere) move. This causes alternate reinforcement and reduction in total signal strength.

a sun spot will greatly increase the number of charged particles in our atmosphere, and you will see interference on a TV screen resulting from the reception of distant transmitters.

At the demodulator of any radio, the effect of the diode is not just to extract the audio signal; there is also a steady voltage present. This voltage is obtained from the effect of the diode on the RF or IF signal, and it is steady only while the strength of the incoming signal is steady; it is large for a strong signal and small for a weak one. The remedy for varying signal strength, then, is to use the steady voltage at the demodulator and feed it back to the IF amplifier stages. The valves that were used for the IF amplifier and (usually) for the mixer stage were of a specialised type (called variable mu) in which the amount of amplification changes when the value of steady voltage applied to an input is changed. By making this feedback connection, the valves could be made to work at full gain when the signal strength was low, but at reduced gain when the signal was large. Using AGC, the receiver could cope with the changes and avoid fluctuations due to shifting reflections.

The mixing of waves is called a heterodyne action, and the oscillator operates at a frequency that is supersonic, meaning higher than frequencies that we can hear (we started to use it to mean faster than sound much later). By the later 1930s you could hardly hold your head high in polite society unless your radio was a superhet. By 1939 the more elaborate radios would use eight valves. One would be used to amplify the radio frequency and followed by a mixer stage to obtain the intermediate frequency, with two more stages of amplification for the IF signals. Following the demodulator there would typically be four valves used for the audio (sound frequency) signals, two of them used to drive the loudspeaker. At the bottom end of the scale, you could buy three-valve radios, with a single valve carrying out oscillator and mixer actions, one IF stage and a combined demodulator and audio output valve. All these valve counts would be increased by one for a set designed to be run from the AC mains, with this additional valve rectifying the AC to a one-way voltage and with a large capacitor added to filter out the remaining AC and smooth the fluctuating voltage.

In the late 1950s, transistors started to replace valves, but the block diagram remains exactly the same because the superhet principle has never been superseded for this type of radio use. The main change in the years from 1960 to the present day has been the increase in the use of FM radios, but the changes that this involved do not appear on the block diagram and consist mainly of using a different type of demodulator; the radios are still superhet types. The AGC principle could be applied even more easily to transistors than to valves, but FM radios use, in addition, another type of automatic control, automatic frequency control (AFC).

FM radio makes use of carriers in the higher radio frequencies in the 80 to 110 MHz range, and an IF frequency of 10.7 MHz, and it's more difficult to make oscillator circuits that will produce an unchanging frequency in this range. Oscillators suffer from drift, meaning that as temperature changes, the oscillator frequency also changes. The percentage change might be small, and for an AM receiver working at 1 MHz, a 0.1% drift is only 1 kHz and not too noticeable, but for an FM oscillator working at 100 MHz a 0.1% drift is 100 kHz, totally out of tune. The FM signal needs a wider bandwidth than AM, and usually up to 230 kHz is allowed.

AFC uses another steady voltage that is generated in an FM demodulator and which depends on the frequency of the signals. By using this voltage to control the frequency of the oscillator (a *voltage controlled oscillator* or VCO), the FM receiver could be locked on to a signal and would stay in tune even if the oscillator components changed value as they changed temperature. Modern FM receivers have very efficient AFC which is usually combined with a muting action so that there is no sound output unless you are perfectly tuned to a transmitter. As you alter the tuning, each station comes in with a slight plop rather than with the rushing sound (or noise) that was so familiar with the earlier FM radios.

The main changes since the 1970s have been the replacement of transistors by ICs so that modern radios contain very little circuitry and most of the space is used for the battery and the loudspeaker. The circuitry is almost identical on all of them, and often is made in the same factories which can be anywhere in the world. As we know so well, the name on a radio is no guide to where it was manufactured, and manufacturing in the UK virtually came to an end in the 1960s when a misguided attempt by the Wilson government to protect the electronics industry made it cheaper to import than to manufacture.

Summary

The superhet principle was one of the most important in radio history, and is still in use. The incoming radio frequency is converted to a lower, fixed, intermediate frequency, and this lower frequency is amplified and demodulated. Because most of the amplification is at this lower fixed frequency, there is much less likelihood of problems arising from feedback of this frequency to the input of the receiver, and both sensitivity and selectivity can be improved. The use of automatic gain control reduces the effects of varying strength of received signals.

Stereo radio

Stereo sound from discs and tape was well established by the 1950s, but stereo broadcasting took a little longer. Stereo broadcasting at its simplest can be achieved by using two separate transmitters, one for each channel,

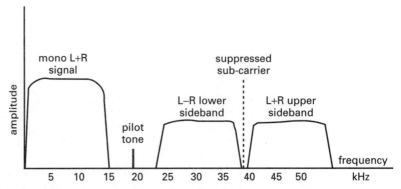

Figure 6.11 The spectrum (graph of amplitude plotted against frequency) for a stereo FM radio signal. This shows the pilot tone and the sidebands of the sub-carrier which are added to the mono signal. This mixture of frequencies is modulated on to the main carrier at around 100 MHz.

but this is wasteful and the stereo effect can be disturbed if the two frequencies are subject to interference at different times or to different extents. Though the first experiments in stereo broadcasting used separate frequencies, it was obvious that stereo broadcasts would not succeed unless two conditions were met. One was that a single frequency must be used and the other was that users of mono receivers should still be able to pick up a mono signal when they tuned to a stereo broadcast. As it happened, these were very similar to the conditions that had been imposed on colour TV systems, see Chapter 8, in 1952, and similar methods could be adopted.

The system that evolved in the USA and which was adopted with only a few modifications in the UK was one that used a *sub-carrier*. As the name suggests, this is a carrier which can be modulated, and which is then itself used to modulate the main carrier. The other important step was how to use the left (L) and right (R) channel information. The solution was to modulate the main carrier with the sum of the channel signals, L + R, which allowed any receiver tuned to the correct frequency to pick up this signal, which is a mono signal. The stereo information was then contained in a signal which was the difference between the channels, L – R. This difference signal has a much smaller amplitude than the L + R signal and is needed only by stereo receivers, making it suitable for modulating on to a sub-carrier.

Figure 6.11 shows the frequency bandwidths of a stereo transmission before modulation on to the main carrier. The bandwidth that will be needed by the mono signal (L + R) is 15 kHz, and whatever other frequencies we use must not overlap this set. A single low-amplitude 19 kHz signal is also present – this is called the *pilot tone* and is used at the receiver, see later. The L – R signals are *amplitude* modulated on to a 38 kHz subcarrier, and this subcarrier is then removed (a *suppressed carrier*), leaving only the sidebands of the modulated signal. These also need a bandwidth of about 15 kHz on each side of the 38 kHz mark. This set of different signals in different frequency ranges is used to frequency-modulate the main carrier of around 100 MHz.

Why remove the sub-carrier and supply a 19 kHz signal? The sub-carrier contributes nothing to the signal; it carries no information. It does, however, require transmitter power, and removing the sub-carrier makes an appreciable saving, allowing the transmitter to carry more of the useful sideband signals. The L – R signals cannot easily be demodulated, however,

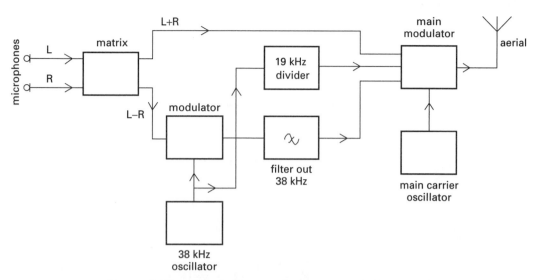

Figure 6.12 A block diagram for the stereo FM transmitter. The L and R signals are mixed to produce the mono signal L + R and the smaller difference signal L – R which is modulated on to the sub-carrier.

without supplying a 38 kHz signal, and this can be supplied locally at the receiver providing there is a small synchronising signal available. This is the purpose of the 19 kHz pilot tone, the frequency of which can be doubled to 38 kHz and used to correct the phase and frequency of the local oscillator in the receiver. The transmitter power used for the 19 kHz tone can be small, because the amplitude of this signal is small. As a bonus, the 19 kHz wave can be used at the receiver to turn on an indicator that shows that a stereo signal is being received.

Figure 6.12 shows the block diagram for transmission. The separate L and R channel signals are added and subtracted in a circuit called a matrix to give the L + R and L – R signals, and the L + R signal is used to frequency-modulate the main carrier. A master 38 kHz oscillator is used to provide the carrier for the L – R signals, and the 38 kHz signal is also used to provide a 19 kHz signal for the pilot tone which is also frequency-modulated on to the main carrier. The modulated L – R signal is passed through a filter which removes the 38 kHz sub-carrier, so that the sidebands can be frequency-modulated on to the main carrier. The modulated main carrier is then amplified and used to supply the transmitting aerial.

Figure 6.13 is the block diagram of the stages following the demodulator for a stereo FM receiver. The early stages of the receiver are exactly the same as they would be for a mono FM receiver, right up to the point where the FM signal is demodulated, providing three sets of signals. A low-pass filter separates off the L + R main signal. A resonant circuit separates out the 19 kHz pilot tone and this is used to control the phase and frequency of a 38 kHz oscillator. Finally, a high-pass filter separates out the sidebands of the L – R signals.

A circuit called a phase-sensitive demodulator has inputs of the 38 kHz carrier frequency (obtained by doubling the 19 kHz pilot tone or by using it to synchronise AM oscillator) and the L – R sidebands, and its output is the L – R signal itself. The L + R and L – R signals are fed into another matrix circuit, producing the L and R signals at the outputs. If you wonder how this is done, think of what happens when you add the inputs:

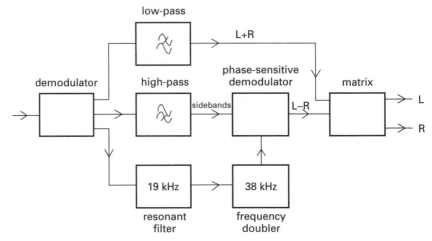

Figure 6.13 A block diagram for a stereo FM receiver. Only the portions from the FM demodulator are shown, as the remainder is a standard superhet block diagram.

$$(L + R) + (L - R) = 2L$$

and also subtract them:

$$(L + R) - (L - R) = 2R$$

so providing the separate L and R signals.

Snags? The stereo signal is more easily upset by interference, particularly car-ignition and other spark interference. A mono signal can use a low-pass filter to remove much of the effects, but this would remove the L – R signals from a stereo transmission. In addition, the noise level of a stereo transmission is always higher because the bandwidth is greater, and noise depends on bandwidth, the only noise-free signals are those with zero bandwidth, and signals with no bandwidth carry no information.

Stereo FM radio that is compatible with mono broadcasts is made possible by the use of a sub-carrier. The main signal consists of the sum of L and R channel audio signals, and this is a mono signal. The difference signal, L – R is much smaller and is modulated on to a 38 kHz sub-carrier, and this carrier is filtered out, leaving only the sidebands. A 19 kHz pilot tone is added as a way of synchronising a local oscillator at the receiver. All three signals are frequency-modulated on to the main carrier. At the receiver, the L + R signal can be used as a mono signal, and if stereo is required, the pilot tone has its frequency doubled and is used to synchronise the phase of a 38 kHz oscillator. This along with the sidebands of the sub-carrier can be used to demodulate the L – R signal, and combining the L + R and L – R signals provided separated L and R audio channel signals. The pilot tone can be used also to switch a stereo indicator lamp on or off.

Chapter 7   Disc and tape recording

Electric gramophones

The first gramophones were totally mechanical. The sound that was recorded was caught in a horn and used to vibrate a membrane that in turn moved a stylus. This stylus scratched a wavy line on a revolving wax cylinder, driven by clockwork, and the system was later applied to wax discs. These discs could be reproduced by electroplating and printing to make shellac discs with the same imprint of a wave. Playing a record was simply the reverse of this process, spinning the shellac disc at a steady speed, eventually standardised at 78 revolutions per minute, and picking up the sound using a needle in the groove the movement of which vibrated a membrane at the end of a horn. Gramophones like this featured in many households in later Victorian times and were still in production, particularly in portable form, in the 1930s. Edison's original idea of 1877 had a long life.

The use of valves for amplification promised to solve one of the problems of the acoustic gramophone, which was the lack of any effective volume control. You could always reduce the volume by stuffing a few socks down the horn, but increasing it was out of the question, though a few inventors harnessed compressed air to make amplifying horns. The main focus for improvements, however, was the recording process. Trying to get a full orchestra around a brass horn was difficult, and no form of control was possible. The possibility of using more than one microphone, being able to control treble and bass response with filters and to ensure that the record tracks did not overlap (through overloading) spurred the development of electrical recording methods. By the early 1930s, virtually all records were being made using electrical methods and the systems that were developed also contributed to the addition of sound to silent films in the late 1920s.

Figure 7.1 shows a typical electrical recording system, and this block diagram is applicable right up to the time when compact discs (CDs) started to replace the older vinyl (LP) records. In the block diagram, several microphones are shown connected to a mixer, so that some control can be obtained, allowing the sound balance of a soloist and of different sections of an orchestra to be achieved. This type of system also makes it possible to reduce background noise, like the coughs and sneezes of an audience, to some extent. This mixer stage is followed by amplification and then by filtering and control stages for bass and treble, then by a driver (a power amplifier). The signals from the drive operate the current head which as the name suggests cuts a track in the master disc which can be metal, wax or plastic. The power that is needed for a good-quality cutter can be very large, and in the latter days of LP discs cutter heads might require up to 1 kW of power.

The bass and treble filters need some explanation. A sound source like an orchestra has an immense range of amplitude (typically 100 dB) from its softest to its loudest, and this range simply cannot be recorded on to a disc or a tape. Taking the disc example, the softest sounds simply leave so little trace that only the noise of the needle on the disc can be heard. The loudest

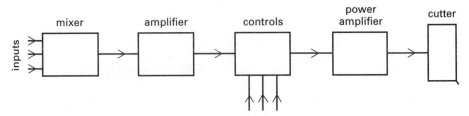

Figure 7.1 Principles of disc recording. The signal is amplified and filtered and applied to a power amplifier which drives the cutter head. This cuts a master disc which is then used to produce other masters for pressing copies in vinyl plastic.

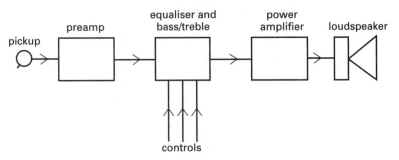

Figure 7.2 Disc replay. The feeble signals (often less than 1 mV) from the pick-up are amplified, and then equalised. At this stage, tone and volume controls can be used. The power amplifier then provides the drive for the loudspeaker.

sounds over-drive the cutter so that the waves on one part of the track overlap and break into an adjacent part of the track, making the record unplayable. The solution is to decrease the range, boosting the softest sounds, and reducing the loudest. In addition, even if the range of amplitudes is fairly small, the treble has to be boosted because most of the noise of a disc is in the form of a high-pitched hiss which would otherwise drown out the highest notes of music, and the bass has to be reduced because it is more likely to cause track overlapping.

The record player then has to reverse these processes, boosting the bass frequencies (typically 30 Hz to 120 Hz) and reducing treble (typically 4 kHz to 10 kHz), and this process, called *equalising*, is possible only if electronic methods are used; you can't use old socks for this task. In addition to equalising, variable controls for treble and bass are needed to adjust the sound to compensate for the size, shape and furnishing of the room in which it is heard. Figure 7.2 shows a block diagram for a record player of good quality with separate bass and treble controls. The signal is obtained from a pick-up, another type of transducer which is similar in construction to a microphone, but using the vibrations from a stylus on the record groove rather than from a membrane.

Any pick-up that is of reasonable quality suffers from a very small signal, in the region of a few millivolts at most, so that a fair amount of amplification is needed. The equalisation, volume, bass and treble controls have to be applied at a stage where the amplitude of the signals is reasonably large, a volt or so, and the final portion is power amplification to drive a loudspeaker. This block diagram is, once again, one that has remained much the same from the 1930s to the 1980s, because the change from the older shellac

78 r.p.m. record to the $33\frac{1}{3}$ r.p.m. long-playing vinyl disc required no change in the methods for a single channel (mono) signal. As for radio, changing first from valves to transistors and then to ICs made no difference to what was being done, only to the details of how it was achieved.

The system of disc recording pioneered by Edison uses a groove cut in a shellac or vinyl disc and modulated by the sound waves. For good quality recording, the bass amplitude must be reduced and the treble frequencies boosted on recording and the opposite actions carried out on replay. Electrical recording and reproduction offers much more control over the process and has been used almost universally since the 1930s.

Hi-fi and stereo

In the later 1940s the word 'Hi-fi', an abbreviation of 'high fidelity', burst in from the USA, and brought new life to the old gramophone. The spirit of hi-fi was that the sound which you heard at home should be as good as it was in the studio or concert-hall. The ideal was always unattainable, but the principle was good – the quality of most gramophones at that time was abysmal, as bad as the quality we now put up with from small transistor radios and from loudspeakers in almost every shop. The nearest anyone ever got to reasonable quality at that time was in some cinemas (when they played music from a disc before or after the film) and good sound was a feature of the original version of Disney's *Fantasia* which could be shown in full glory only in cinemas which were equipped for stereo sound.

Hi-fi required large loudspeakers (*woofers*) that could reproduce the bass notes along with small ones (*tweeters*) to do justice to the higher notes. It also required good pick-ups, good preamplifiers, separate treble and bass controls and, above all, an excellent power amplifier stage. All of these requirements could be satisfied by good designs using valves, and the amplifier that set the standard in the 1950s was the Leak, with its (then) remarkable figure of less than 0.1% distortion. In these days, home construction was popular, and many firms that became well-known started with a home-made system which then was put into quantity production. Some of the amplifiers of that time are now valuable collectors' items, even the kit models that were sold for home assembly.

Hi-fi was about quality, and this cannot be indicated in a block diagram, so that what we have seen in Figure 7.2 still holds. The development of stereo sound, however, made some changes. The principle was not new, and the idea of supplying each ear with a slightly different sound pattern had been established as long ago as the start of the twentieth century. This creates an impression that the sound is no longer coming from one small space, and all the first listeners to stereo talk about a 'feeling of space', a 'broad band of music' and so on. What was lacking in the early days, however, was any way of achieving stereo sound on a standard gramophone disc, though the principle that was adopted was based on one patented in 1936 by A. D. Blumlein (whose list of patents covers almost everything we think of as modern).

Disc stereo recording uses a cutting head whose stylus can be moved in two directions that are at right angles to each other, cutting tracks in the walls of the groove. With the signals from two separate (left and right) microphones used to control the currents, one wall of the groove contains a left track and the other a right track, Figure 7.3. Replay also uses a single stylus that is connected to two pick-up elements, each sensing movement in one of two directions at right angles, so that the outputs are of left and right signals.

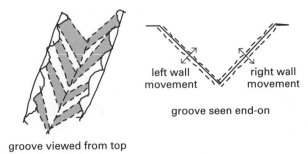

groove viewed from top

left wall
movement

right wall
movement

groove seen end-on

Figure 7.3 Stereo disc recording principles. The walls of the groove are at 90° to each other, and they are separately modulated with L or R signals. The principle is that two mechanical motions at 90° to each other do not interfere.

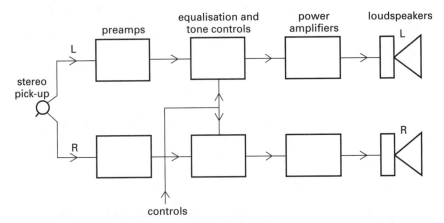

Figure 7.4 A block diagram for a stereo disc replay system. This consists of two separate amplifiers, sharing a power supply, and with volume and tone controls ganged so as to operate together. There must also be a balance control to allow for loudspeaker positioning and other factors which might bias the sound to one side or the other.

The stereo gramophone block diagram looks as Figure 7.4, with separate L and R signals from the pick-up passing through their own independent amplifier chains. Each channel (L or R) has treble and bass controls, and these are normally ganged. The volume controls can be separate, but it is more usual to gang them also and use an additional *balance* control to set the ratio of left to right volume. The balance control is normally set so that the main sound appears to come from between the two loudspeakers that are arranged one to the left and one to the right of the listener(s).

The coming of stereo caused a momentary halt in the hi-fi process, because listeners were more impressed by stereo of any kind than by high-quality single-channel sound (and this has not changed appreciably). Another factor that has limited the appeal of high-quality sound has been the emergence of pundits who have brought to sound reproduction the words and the mystique of wine tasting (but often with little justification). The plain truth is that it is easier now to achieve good-quality sound than it has ever been in the past, and without elaborate and costly equipment. Achieving the best quality is, as always, quite another matter, particularly if the reproduction of sound depends on a hard diamond stylus scraping along a piece of slippery plastic.

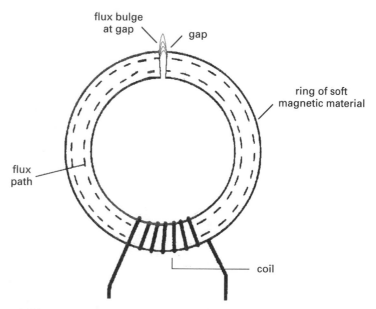

Figure 7.5 A primitive tape-head in detail. The alternating current through the coil causes a strong alternating magnetic field at the gap which will magnetise any magnetic material that is in contact with the gap.

The popularity of hi-fi as a hobby in the late 1940s led to an interest in better standards of reproduction, boosted by the vinyl LP record and the use of stereo recording. Hi-fi required good pick-ups, well-manufactured records, good amplifiers and good loudspeakers and though the transition from valves to transistors (and to ICs) was slower than it was for other branches of electronics, it eventually happened. Valve amplifiers are, however, still available.

Tape and cassette

We often think of tape recording as comparatively new, but it was invented by the Dane Valdemar Poulsen in 1889, and a recording that was made in 1900 by Emperor Franz Josef of Austria-Hungary has survived. The principle, Figure 7.5, is simple enough. A ring of magnetic material, with a coil of wire wound round it, will be magnetised when current flows in the coil. If there is a small gap in the rim, then the magnetisation around this gap will be very strong, and when the current through the coil is AC, the magnetisation around the gap will also alternate in direction and strength. This concentrated magnetic field will affect any magnetic material near it.

Sound is converted, using a microphone, into electrical signals, and these are connected to the coil. A magnetic material, which originally was a steel wire, is drawn past the magnetic core at a steady speed, so that each portion of the wire will be magnetised to a different extent, depending on how much current was flowing in the coils when that portion of wire was drawn past the core. To replay this, the process is reversed, connecting the coil to headphones. As the wire is drawn past the core, the changing magnetism causes currents to be created in the coil and these are heard as sounds in the headphones.

Poulsen's Telegraphone, as he called it, worked and was easier to use than a disc recorder, but it never became a home item, unlike Edison's phonograph.

The ability to record was not something that was important for entertainment in the days before radio, but the main point that made the wire recorder unattractive for the home was the need to use headphones – only one person could hear the faint recorded sounds, whereas the sound from a disc recording could fill a room with the 'full, rich notes' from its horn (the description dates from 1897). Poulsen's invention was used to a limited extent as an office dictating machine up to the 1920s but it was superseded by later disc-based recorders.

The principle never totally died out, however, and radio companies experimented with wire recorders as a way of preserving important performances. The record companies also took an interest, because recordings on wire could be edited, unlike recording made on wax discs, and by using enough wire, a recording lasting 20 minutes or more could be captured. This was a significant advantage, because a wax disc could hold only about five minutes of recording. The limitations caused by the use of wire, however, were difficult to deal with, and these old recorders used (literally) miles of wire moving at high speed across the magnet core or recording head.

The rebirth of magnetic recording occurred in Germany during the 1939–45 war. The BASF company, famous for aniline dyes, developed plastic tape with a magnetic iron-oxide coating that was immensely superior to iron or steel wire for recording uses, and could be used at much lower speeds of around 15 inches per second. Tape recorders of advanced design were found when the Allies invaded Germany, and were passed to electronics companies in the UK and USA for inspection. As a result, tape recorders became commercially available in the 1950s. In this respect, electronics firms were more commercially aware than the British motor industry who were offered free manufacturing rights on the Volkswagen beetle, but rejected it on the grounds that such a curious vehicle could not be a commercial proposition.

The record companies and radio companies were delighted with these developments, and they started a demand for high-quality tape equipment which still exists. Apart from the editing convenience, the use of tape allowed for stereo recording (by using two tracks recorded on the tape) at a time when this was very difficult to achieve on disc, and also for sound effects that had not been possible earlier. The use of tape recorders at this level fed down and resulted in an interest in sound recording at home.

Figure 7.6 shows a typical block diagram for a domestic tape recorder. The tape-head is a refined version of the coil and core arrangement, using a tiny gap in a magnetic metal to form a concentrated magnetic field across which the tape is moved during recording. The same head is used for replaying, and a switch allows the head to be connected to the record or replay circuits. These blocks contain familiar portions, but the bias/erase oscillator block requires some explanation.

The problem of magnetic recording is that magnetic tape is not a linear medium – a graph of recording current through the coil of the recording head plotted against the stored magnetism is not a straight line, more like an 'S' shape, with no retained magnetism at all for small currents, Figure 7.7. This makes the replayed sound appear impossibly distorted, even for speech. Early on, this had been tackled by adding a permanent magnet near the head, which improved matters but not to such an extent that tape could be considered good enough for recording music. The problem was solved by adding a *bias wave* at a high frequency of around 50 to 100 kHz. When this is done, using an amplitude that has to be carefully set for the type of tape and the tape speed, the results can be acceptable. With meticulous adjustment and good circuit design, tape recording could be good enough to

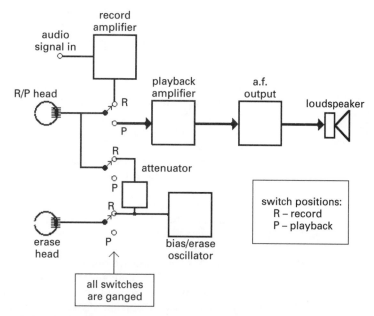

Figure 7.6 A block diagram for a tape recorder, showing the record/replay switching. As is normal, the same head is used for both recording and replay, though better results can be obtained by using three separate heads (record, erase and replay).

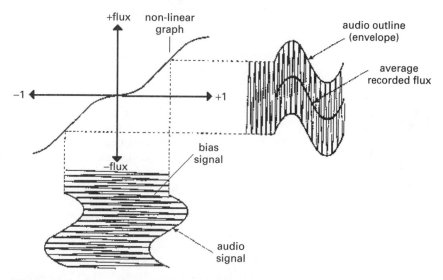

Figure 7.7 The effect of bias with a high-frequency signal is to overcome to some extent the non-linear shape of the graph of magnetisation plotted against signal strength.

making the master recordings for gramophone records, and was used in this way from the 1940s onwards. Making the quality acceptable on home recorders which had to be produced at a reasonable price was another matter.

The upper limit of the range of frequencies that can be recorded is affected by the tape speed and the head gap. To get a good response for high frequencies demands either a very small head gap (a gap of 0.001 inches is an enormous gap!) or fast tape speeds. The other problem is noise, because the nature of magnetism makes tape a noisy medium. Noise is less of a worry on wide tape, but the trend has been to use narrower tape, often with four tracks so that the track widths are very narrow.

As it happened, the novelty wore off, mainly because the tape, contained on open reels, had to be threaded into the recorder, and because there was no market in recordings on tape to match the range and variety of gramophone discs. A generation of open-reel tape recorders was eventually scrapped when Philips invented and marketed the compact cassette in 1961. The principle is the same and so the block diagram is the same, but what has improved is convenience. Cassettes originally offered low-quality sound, using narrow tape used at a low speed, and the original intention was to use the machines as dictaphones. The convenience of the cassette, however, led to recorders being marketed both for home recording and also for replaying pre-recorded cassettes, and in the following 20 years remarkable strides were made in improving tape, methods and mechanisms until it was possible in the 1980s to claim with justification that cassette tapes could be used as part of a hi-fi system. Even the introduction of digital audio tape (DAT) has made little impression on compact cassette sales, and all recordings that were available on other media (vinyl disc or CD) are also available on compact cassette.

Tape and cassette recording of sound is not new, but the methods that are used to make the sound of acceptable quality are of recent origin. All magnetic recording makes use of a magnetic material being moved past a recording/replay head which consists of a metal core with a narrow gap and a coil of wire, and the construction of this head is very exacting if good quality recording is required. Though open-reel tape recording is obsolete for domestic use, cassettes have become a standardised and accepted way of distributing recorded music.

Cinema sound

In the twentieth century, no electronic system before TV made such an impression as cinema sound, films with a sound track, in 1928. The successful system, called Phonofilm, was invented by Lee De Forest, who also invented the first amplifying radio valve, and the principles did not change until the use of magnetic tape tracks in more recent times. Early systems had suffered from inadequate volume and from synchronisation problems, but De Forest solved both by using the film itself to record the sound and by making use of valves for amplification.

The block diagrams for a cinema sound system of the traditional type are illustrated in Figure 7.8. On recording, the sounds are converted into electrical signals using microphones, and mixing is carried out as required. The amplified signal is used to drive a 'light-valve', which opens or closes

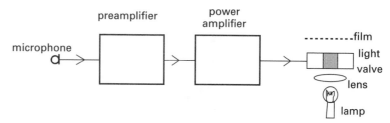

Figure 7.8 Recording cinema sound, using the film to carry the sound track. This depends critically on the design of the light-valve.

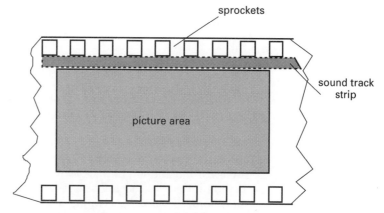

Figure 7.9 The sound track on a 35 mm film.

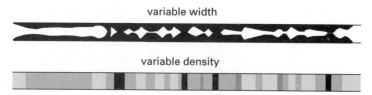

Figure 7.10 Variable width and variable density sound tracks compared. The replay system will cope with either type of track.

according to the amplitude of the signal wave at its input. These light-valves were originally electromagnetic, though other principles have been used. By placing the light-valve near the edge of the film, illuminated by a light beam, a strip of otherwise unused film can be exposed to light that is controlled by the light-valve, creating a wave pattern on this strip of film, the sound track.

Since these early days, two separate principles have been used. One has used the light-valve to vary the *width* of the light beam that reached the film, the other used the valve to control the *brightness* of the light. Whichever method was used on recording, the developed and printed film contained a strip of sound track, Figure 7.9, situated between the frame edge and the sprocket holes. This strip of film has been exposed so that it contains a picture of the sound in the form of variations of either width or of blackness, Figure 7.10. This can be played back using a *photocell*, a type of transducer that responds to light either by generating a voltage or by allowing

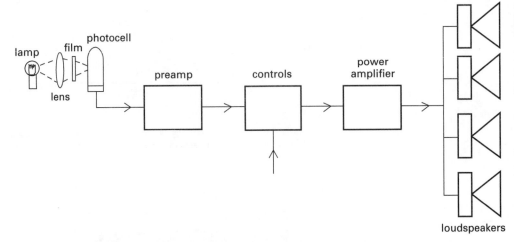

Figure 7.11 The replay arrangement uses a photocell where the (feeble) output signal is amplified. Cinema sound requires large amplifiers to provide power to a bank of loudspeakers.

current to pass from a voltage supply. On playback, the light of the cinema projector hits the sound track as well as the main frames of the film, so that as the film moves the amount of light passing through the sound track to the photocell varies, generating a variable voltage or current which is an electrical signal that can be amplified, Figure 7.11.

Incidentally, the sound track that corresponds to any particular frame of a file is not placed next to that frame. This is because film in a projector or camera does not move smoothly but in jumps. Each frame is held for about $\frac{1}{24}$ second for exposure and then moved rapidly, and this would put a 24 Hz hum on to the sound track if we read the sound track at the point of projection of the frame. Instead, the sound track is placed about 30 frames ahead of the picture at a point where the film is moving steadily rather than in jumps. This solves the hum problem, but it makes editing more complicated than it was for silent films, and it also causes problems if a film breaks and has to be repaired by taking some frames out, because the sound track that is removed does not match the removed picture. Many films now use a 6-track magnetic tape for sound.

Noise

Noise is one of the persistent problems for any type of recorded sound, as we have indicated already. A really old recording on shellac disc will consist so much of hissing and scraping that you wonder how anyone could listen to it with any enjoyment, and though the noise level of discs had been reduced by the 1940s it was the main factor in the popularity of the long-playing vinyl disc which replaced the older type. The lower speed of the LP, along with the use of a long-life diamond or sapphire stylus and equalisation circuits all kept the noise level of the LP reasonably low, and this seemed satisfactory until we heard (or didn't hear) the noise level of CDs.

The noise level on early cassette tapes almost rivalled the old discs. Tape noise is always a problem, and it was dealt with on professional equipment by using wide tape. The width of a cassette tape allocated for four tracks (two in each direction) is only $\frac{1}{8}$ inch, whereas an open-reel machine will use at worst two tracks on $\frac{1}{4}$-inch tape, or use much wider tape such as the

$\frac{1}{2}$-inch tape (now used for video) or more. In addition, the frequency range that can be recorded on tape is very limited when the tape speed is low. Cassette tape moves slowly, at about $1\frac{7}{8}$ inches per second (ips) rather than the 15 or 30 ips of professional tape equipment, and this restricts the highest frequencies that can be replayed to as low as 6 kHz or less. This inability to record or replay the higher frequencies is also controlled by the width of the gap in the tape-head – and a large gap means one thousandth of an inch.

Development of better tape materials has improved the noise performance of tape, and we can now make tape-heads with gaps that are measured in microns, millionths of a metre. Even with all that, however, tape could not be considered suitable for serious use without noise reduction circuitry. Several schemes have been used in the past, but the only ones that have survived have used equalisation methods, deliberately altering the signals before recording and reversing the action on replay. The best-known of these noise reduction systems is Dolby, named after the British engineer Ray Dolby (who at one time also worked in the development of video recording at Ampex Inc.). The Dolby B noise-reduction system is almost universally used for pre-recorded cassettes, and other versions such as C or HX Pro are used on more expensive equipment. There is also another system, dbx, which is astonishingly effective, but has been used on only a few cassette machines (mainly by Technics) and on some radio transmissions in Japan and the USA.

Figure 7.12 illustrates the principle. Tape noise is at a low amplitude and it is concentrated in the higher frequencies. The amplitude of the noise, however, is almost the same as the amplitude of the softest music, and the hiss that you hear when you turn up the volume to hear better is very noticeable because the human ear is particularly sensitive to this range of frequencies. In addition to the noise problem, tape will overload if too large an amplitude of signal is recorded.

On recording, the Dolby circuits split the electrical signal into separate channels. The lower frequency signals pass unchanged. The low-amplitude signals in the higher frequency range are boosted, making their amplitude greater than they were in the original signal. On the C and HX Pro systems, signals that have an amplitude that would cause overloading are reduced, so the result is a signal where the smallest amplitudes are still above the noise level and the maximum amplitude does not overload the tape. When the Dolby recorded tape is replayed, the processing that has been carried out on recording is reversed. The attenuation of the signals that have been boosted will also reduce the noise signals from the tape so that they are almost undetectable and if the more advanced systems have been used, the tape will also deliver sound that has a much greater range (from softest to loudest) than is possible when no processing has been used. Though the C and HX Pro systems offer better noise reduction, the Dolby B system is almost universal on prerecorded cassettes because it does not alter the sound too much. This means that if you replay a Dolby B recording on a player which is not equipped with Dolby decoding the sound does not sound distorted; just a little biased to treble. This is not something that could be said for the other systems (though undecoded dbx can be acceptable in a car because the volume is almost constant).

Due to Dolby and dbx, cassette tapes can be used as part of a hi-fi system, and cassette tapes that have been recorded using systems such as Dolby C, Dolby HX Pro or dbx can rival a CD for low noise and sound amplitude range. Dolby noise reduction is also used extensively in cinema sound, whether optical or magnetic.

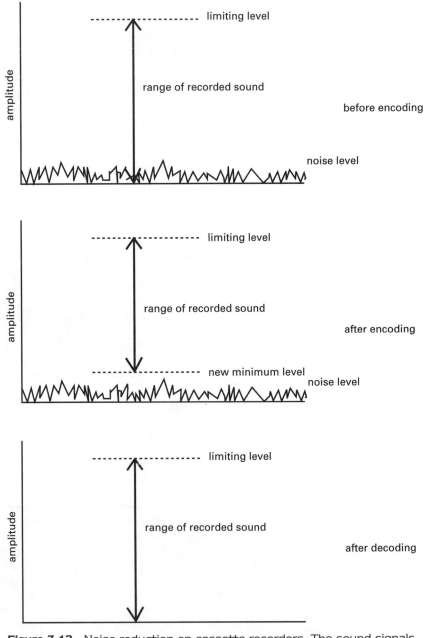

Figure 7.12 Noise reduction on cassette recorders. The sound signals are processed so as to reduce the range, in particular to ensure that no signals fall to a level as low as that of the noise. When the signal together with noise is expanded, the noise disappears.

Summary

Noise is the enemy of all recording and broadcasting systems. Tape noise is particularly objectionable because it consists of a hiss, a type of sound that human ears are particularly sensitive to. For professional recording, the use of wide tape overcomes the problem to a considerable extent because a wider tape has a larger magnetised area and so produces a larger signal that 'swamps' the noise. The narrow tape used on cassettes, however, has a high noise level that cannot be overcome completely by using better tape materials. Noise-reduction systems operate on the principle of boosting the amplitude of the recorded sound when it is recorded and decreasing it on replay, and a particularly effective method of doing this concentrates the boosting and reducing actions on the frequency range that is most affected.

Video and digital recording

The recording of sound on tape presented difficulties enough, and at one time the recording of video signals with a bandwidth of up to 5.5 MHz, and of digital sound, would have appeared impossible. The main problem is the speed of the tape. For high-quality sound recording, a tape speed of 15 ips was once regarded as the absolute minimum that could be used for a bandwidth of 30 Hz to about 15 kHz. Improvements in tape and recording head technology has made it possible to achieve this bandwidth with speeds of around 1 ips, but there is still a large gap between this performance and what is required for video or for digital sound recording. This amounts to requiring a speed increase of some 300 times the speed required for audio recording. Early video recorders in the 1950s used tape speeds as high as 360 ips along with very large reels of tape.

Video recording, even now, does not cope with the full bandwidth of a video signal, and various methods of coding the signal are used to reduce the bandwidth that is required. In addition, the luminance (black and white) video signals are frequency-modulated on to a carrier, and the colour signals that are already in this form have their carrier frequency shifted – see Chapter 8 for more details of luminance and colour signals.

For domestic video recorders, the maximum bandwidth requirement can be decreased to about 3 MHz without making the picture quality unacceptable, but the main problem that had to be solved was how to achieve a tape speed that would accommodate even this reduced bandwidth. In fact, the frequency of the carrier ranges between 3.8 MHz to 4.8 MHz as it is frequency-modulated to avoid the problems of uneven amplitude when such high frequencies are recorded on tape.

The brilliant solution evolved by Alexander Poniatoff (founder of the AMPEX corporation) was to move the recording head across the tape rather than move the tape over a head. Two (now often four) heads are used, located on the surface of a revolving drum, and the tape is wound round this drum so that the heads follow a slanting path (a *helical scan*) from one edge of the tape to the other, Figure 7.13. The signal is switched from head to head so that it is always applied to the head that is in contact with the tape. At a drum rotation speed of around 1500 rotations per second, this is equivalent to moving the tape past a head at about 5 metres per second.

Though the way that the head and the tape are moved is very different from that used for the older tape recorders, the principles of recording remain unchanged. The block diagram for a video cassette recorder is very different from that of a sound recorder of the older type, but the differences are due to the signal processing that is needed on the video signals rather than to differences in recording principles.

Later sound recorders for very high-quality applications use *digital audio tape* (DAT). This operates by converting the sound into digital codes (see Chapter 9) and recording these (wideband) signals onto tape using a helical scan such as is used on video recorders. The main problem connected with DAT is that the recordings are too perfect. On earlier equipment, successive recording (making a copy or a copy of a copy . . .) results in noticeable degradation of the sound quality, but such copying with DAT equipment causes no detectable degradation even after hundreds of successive copies.

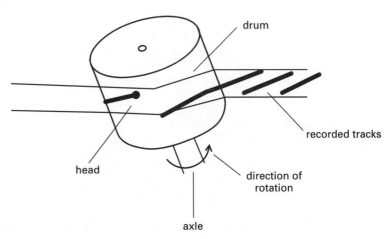

Figure 7.13 Principle of rotary-head video recording. The two (or more) heads are mounted on a drum, and the tape is wrapped at a slight angle. This makes the head trace out a sloping track across the tape as the drum revolves and the tape is pulled around it at a slow rate.

This would make it easy to copy and distribute music taken from CDs, and the record manufacturers have succeeded in preventing this misuse of DAT (though not in the Far East). DAT recorders sold in the UK are therefore fitted with circuits that limit the number of copies that can be made, and this partly accounts for the high price of DAT equipment in the UK as compared to overseas.

Tape as a recording medium hardly seems adequate for sound recording, and its use for video and for digital sound is a triumph of technical development. The main factors have been the use of helical scanning, moving recording/replay heads across the tape rather than moving the tape past a stationary head, and signal processing that reduces the bandwidth required and also makes use of a frequency-modulated carrier which is less affected by problems like frequency response, uneven tape speed and so on.

Scanning

Oddly enough, television has a rather longer history than radio, and if you want to know more of its origins there is a book called *Birth of the Box* (Sigma Press) which deals with the fascinating development of this subject. Almost as soon as the electric telegraph (invented in 1837) allowed Morse code (invented in 1838) messages to be sent along wires, inventors tried to send picture signals, and by the middle of the nineteenth century the principle of *scanning* a picture had been established.

Scanning means breaking an image into small pieces, read in succession, each of which can be coded into an electrical signal.

Suppose, for example, that you drew a picture on a grid pattern such as Figure 8.1. If each square on the grid can be either black or white, you could communicate this picture by voice signals to anyone with an identical grid by numbering each square and saying which squares were white and which were black. It may sound elementary, but it is the whole basis of television and also of fax machines (which are almost identical in principle to the first television transmitters and receivers).

Early television systems used mechanical scanning (with rotating discs or mirrors) and were confined to still pictures in black and white with no shades of grey. Television as we know it had to wait until electronic components had been developed, and in particular, the cathode-ray tube and camera tube. Even by the early 1930s, however, Philo Farnsworth in San Francisco had demonstrated that electronic television was possible, and a form of radar had been used to measure the distance from earth of the reflecting layers of charged particles in the sky (the Heaviside and Appleton layers).

The attraction of the cathode-ray tube is that it can carry out scanning by moving the electron beam from side to side and also from top to bottom, tracing out a set of lines that is called a *raster*. Nothing mechanical can produce such rapid changes of movement as we can obtain by using an electron beam, so that most of the television pioneers realised that this was the only method that would be acceptable, though the cinema industry still believed that mechanical systems would prevail. By the 1930s, it was becoming accepted that any TV receiver would certainly use a CRT, and the main research effort then went into devising a form of CRT that could be used for a TV camera. The principles of such a tube were:

1 a light-sensitive surface on which an image could be projected using a lens, causing electrons to leave the surface;
2 a scanning electron beam which could replace the electrons lost from the light-sensitive surface;

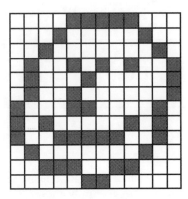

Figure 8.1 A simple image consisting of black or white squares. We can imagine this being built up line by line by scanning across and deciding whether a square should be filled or not.

3 some method of detecting how many electrons were being replaced for each portion of the image (this is the signal current).

Without going into details, a camera CRT must allow the image projected on to the face of the tube to be scanned by an electron beam and it must produce a signal current or voltage that can be amplified so as to represent the brightness of each part of the scanned picture.

Synchronisation

The problems of developing cathode-ray tubes were only part of the story. A remarkable lecture in 1911 by A. A. Campbell-Swinton outlined a television system that we can recognise today and which was at least 20 years ahead of its time. It imagined that both camera and receiver would use cathode-ray tubes, and for the first time emphasised that the scanning had to be synchronised, and that this could be done by using scanning signals that had also to be transmitted so that the scanning at the receiver would remain in step with the scanning at the transmitter. Figure 8.2 shows Campbell Swinton's drawing.

We can make a block diagram, Figure 8.3, for a simple TV system that follows closely along the lines laid down by Campbell Swinton. This shows a camera tube which is scanned by signals from a generator, and these signals are also sent to a transmitter along with the amplified vision (*video*) signals from the tube. These signals can be sent down a single wire (or they can be modulated onto a radio wave) to the receiver where the video signals are separated from the scanning signals, and the separate signals are applied to the receiver CRT. This is virtually a block diagram for the first television systems used from 1936 onwards.

The most remarkable aspect of the development of electronic TV in the 1930s was that so many new problems were tackled at the same time. The TV signal needed a large bandwidth, thousands of times greater than anything that had been used in radio. The carrier frequency had to be high and no radio valves at the time were suitable to be used at such high frequencies. Circuits had to be invented to generate the scanning waveforms, which needed to be of sawtooth shape. A system had to be worked out for allowing the receiver to generate its own sawtooth scanning waveform, but keep these perfectly synchronised with the scanning at the transmitter. Not until

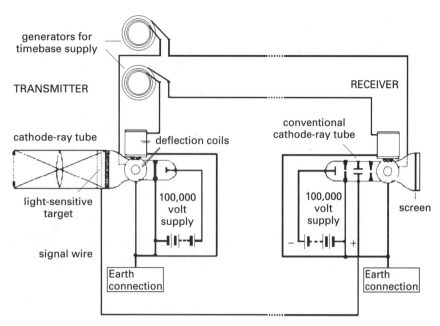

Figure 8.2 A drawing of Campbell Swinton's proposal for an electronic television system.

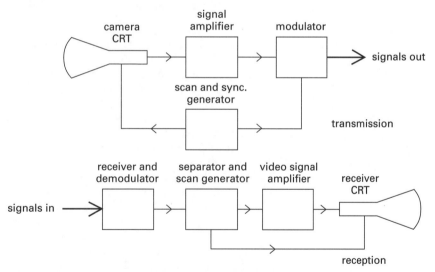

Figure 8.3 A modern version of a simple electronic TV system.

the space race would any team of engineers tackle so many new problems together on one project.

Though primitive television systems used mechanical scanning, television as we know it was impossible until electronic methods of scanning a picture were devised. The scheme outlined by Campbell Swinton was the basis of modern television, and it made use of a form of cathode-ray tube at both transmitter and receiver. This scheme also emphasises the use of synchronising signals that would be used by both transmitting and receiving equipment.

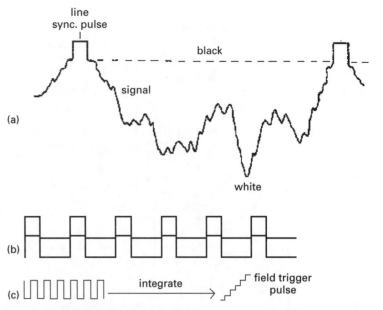

Figure 8.4 The standard TV signal (for a black-and-white picture), showing the signal level for one line and the pulses that are transmitted at the end of a field. The signals are shown as they appear when modulated on to the carrier, with the tips of the sync. pulses representing peak carrier amplitude.

The TV waveform

The solutions to many of the problems are illustrated by looking at the shape of the TV waveform. Like any wave, this is a repeating pattern, but the shape is much more complicated than that of a sound wave. In particular, the wave contains some portions that have very sharp edges, and these edges are vital to synchronisation. A typical single wave for a black-and-white TV system of the 1950s is shown in Figure 8.4a, along with a portion of the waveform that is transmitted at the end of a 'field' when all of a set of lines have been scanned and the beam is at the bottom of the tube face, Figure 8.4b.

The small rectangular portion of the wave in Figure 8.4a is the line synchronising pulse ('line sync.'), and it is used in the receiver to ensure that the electron beam starts scanning just as the first part of the video wave reaches the receiver. The video wave itself will have a shape that depends on the amounts and position of dark and light parts of the picture across one line. Mechanical TV originally used a 30-line scan, but electronic TV started in the UK with 405 lines (525 lines were used in the USA), a huge improvement in resolution that allowed relatively fine detail to be seen. The video amplitude ranges between black and peak white, and the synchronising pulses are in the opposite direction, lower than the level that is used to represent black so that nothing can be seen on the screen when the pulses are transmitted.

At the end of a set of lines, the waveform changes. There is no video signal, just a set of closer-spaced synchronising pulses. These are used at the receiver to generate a *field* synchronising pulse which will bring the beam back to the top left-hand corner of the screen ready to scan another set of lines. As before, these pulses have an amplitude that is lower than black level ('blacker than black') so that they do not cause any visible disturbance on the screen.

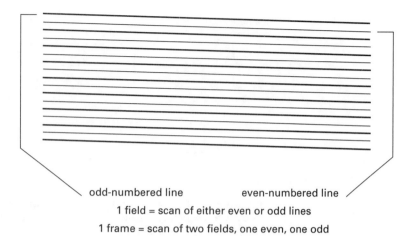

odd-numbered line   even-numbered line

1 field = scan of either even or odd lines

1 frame = scan of two fields, one even, one odd

Figure 8.5 The principle of interlacing. This allows a more detailed picture to be transmitted without using an excessive bandwidth.

As it happens, using a full set of 405 lines to make a picture was out of the question when TV started in 1936 (and it would still cause problems even now). The trouble is that a waveform consisting of 405 lines repeated 50 times per second needed too much bandwidth. Pioneers in the USA had even more to cope with, using 525 lines at 60 pictures per second. The answer that evolved on both sides of the Atlantic was *interlacing*. Interlacing, using the UK waveform as an illustration, consists of drawing the odd-numbered lines in $\frac{1}{50}$ second and the even numbered line of the same picture in the next $\frac{1}{50}$ second, Figure 8.5. This way, the whole picture is drawn in $\frac{1}{25}$ second, and the bandwidth that is needed is only half as much as would be needed without interlacing. Interlacing is still used on TV pictures, though it is not used on computer monitors because on the small bright images of a monitor the use of interlace can cause a flicker which is visually disturbing.

We can look now at a block diagram of the studio and transmission side of black and white TV – we'll look later at the additions that have to be made for colour.

The TV waveform is not a symmetrical wave of fixed shape. The synchronising pulse for each line occurs at regular intervals, but the waveshape that follows this pulse depends on the distribution of light and shade in that line of the picture. The use of interlacing, scanning only half of the total number of lines in each vertical sweep, reduces the bandwidth by half without degrading the picture quality.

Transmission

TV transmission at its simplest starts with a TV camera, which is fed with synchronising pulses from a master generator. Light from the scene that is to be transmitted is focused through a lens on to the face of the camera tube, and this light image is converted into an electron charge image inside the front section of the tube. The scanning electron beam discharges these, and the discharge current is the video signal. This is the starting point for the block diagram of Figure 8.6. The video signal, which is measured in microvolts rather than in millivolts, has to be amplified, and the synchronising pulses are added. The generator that supplies the synchronising pulses

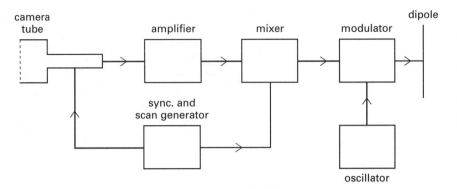

Figure 8.6 TV transmission from camera to aerial. This block diagram shows a simplified version of the transmission of a monochrome picture.

also supplies the scans to the camera tube. The complete or *composite* video signal that contains the synchronising pulses is taken to the modulator where it is amplitude-modulated on to a high frequency carrier wave. Modern TV systems modulate the signal so that peak white is represented by the minimum amplitude of carrier and the synchronising pulses by the maximum amplitude. In the original UK TV system the signals were modulated the other way round, with peak white represented by maximum carrier amplitude, but this method allowed interference to show as white spots on the screen. When the system changed to 625 lines in 1968, maximum carrier amplitude was used for the sync pulse peaks in line with the methods used in other countries.

The sound signal has been omitted from this to avoid complicating the diagram. The sound is frequency-modulated on to a separate carrier at a frequency higher than that of the vision carrier, and is transmitted from the same aerial.

The video signal is generated from a camera tube and is amplified and combined with synchronising signals to form the composite video signal. This is amplitude modulated on to a carrier with the tips of the sync. pulses represented by peak carrier amplitude. The sound signal is frequency modulated on to a separate carrier which is at a frequency 6 MHz higher than the vision signal.

TV receiver

A modern TV receiver for monochrome, Figure 8.7, has to deal with several separating actions. The signal from the aerial is processed in a straightforward superhet type of receiver, and the methods that are used differ only because of the higher frequencies and the larger bandwidth. The vision and sound carriers are received together, and so the IF stages need to have a bandwidth that is typically about 6 MHz, enough to take the wideband video IF and the sound IF. The real differences start at the demodulation block. A typical IF range is 35 to 39.5 MHz (for amplitude 6 dB down from maximum). Figure 8.8 illustrates the IF response which must be broad enough to include both sound and vision carriers and sidebands, yet provide enough filtering to exclude signals from the adjacent frequencies. These adjacent frequencies arise from the mixing of the superhet oscillator frequency with the signals from other transmitters.

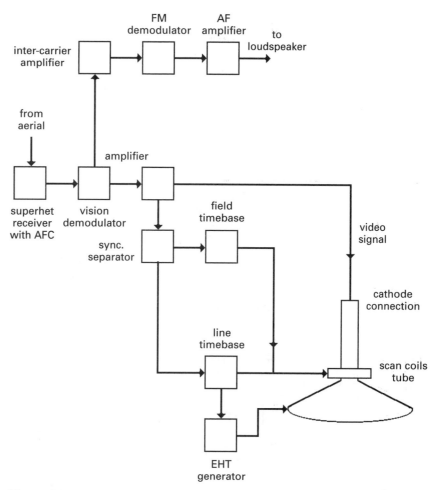

Figure 8.7 A monochrome receiver. The diagram shows all of the superhet circuits as one block, and concentrated on the vision (luminance) and sound (audio) signals.

This drawing also shows the frequency which is used as the carrier for colour signals, see later.

The demodulator is an amplitude demodulator, and at this stage the composite video signal with its sync. pulses, can be recovered. The effect on this stage on the IF for the sound signal is to act as a mixer, and since the sound IF and the vision IF are 6 MHz apart, another output of the demodulator is a frequency-modulated 6 MHz signal (the *intercarrier* signal) which carries the sound. This is separated by a filter, further amplified and (FM) demodulated to provide the sound output.

The video signal is amplified, and the synchronising pulses are separated from it by a selective amplifier. These pulses are processed in a stage (the *sync. separator*) which separates the line and the field pulses by using both differentiating and integrating circuits, Figure 8.9. The differentiated line pulses provide sharp spikes which are ideal for synchronising the line oscillator, and the field pulses build up in the integrator and so provide a field synchronising pulse. The effect of the differentiator on the field pulses is

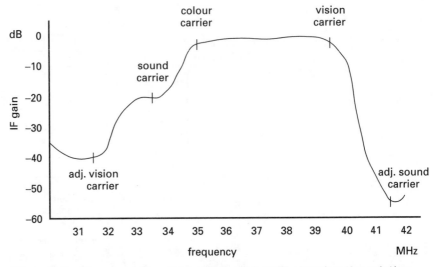

Figure 8.8 The response of a typical receiver IF, showing the relative signal strengths over the bandwidth. The frequency marked as colour carrier is important for colour receivers.

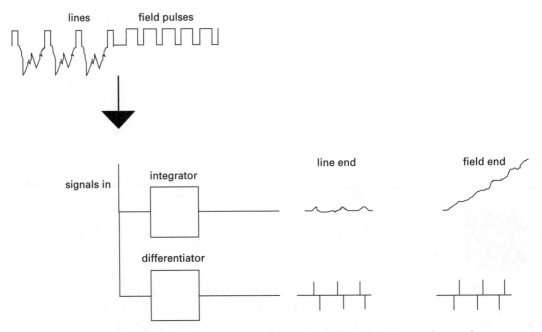

Figure 8.9 How sync. pulses are separated by integrating and differentiating circuits. The output of the differentiator during the field interval is ignored.

ignored by the receiver because the beam is shut off during this time, and the effect of the line pulses on the integrator is negligible because these pulses are too small and too far apart to build into a field pulse.

These pictures of sync. pulses are simplified. At the end of a field or frame, there are five pulses called equalising pulses placed before and following the field sync. pulses. The aim of this is to give time for the circuits to adjust to the differences between the end of a field and the end of a complete frame (there is half-a-line difference in timing).

The synchronised line and field timebase oscillators drive the output stages that deflect the electron beam, using magnetic deflection (passing scan currents through coils). The fast flyback of the line scan current causes a high voltage across the transformer that is used to couple the line scan output to the deflection coils, and this voltage is used to generate the EHT (extra high tension) supply of around 14 to 24 kV that is needed to accelerate the beam.

Meantime, the video signal has been amplified further and is taken to one terminal (cathode) of the CRT, with another terminal (grid) connected to a steady voltage to control brightness. Another set of circuits, the power supply unit (PSU) uses the AC mains supply to generate the steady voltages that will be used by the receiver circuits.

A monochrome receiver uses the normal superhet circuit up to the vision demodulator, where the composite video signal is recovered and the 6 MHz intercarrier sound signal is filtered off, demodulated and amplified. The composite video signal is passed to a sync. separator which removes the video portion and allows the two sets of synchronising pulses to be separated and used to synchronise the timebases. The video signal is applied to the grid (or cathode) of the CRT, and the timebase signals to deflection coils. The very large pulses that exist on the line times base output are stepped up by a transformer and used to generate the EHT supply of 14 kV or more for the CRT.

Colour TV

Though Sam Goldwyn is reputed to have said that he would believe in colour TV when he saw it in black and white, the idea of colour TV is not new, and methods for transmitting in colour have been around for as long as we knew that black-and-white (monochrome) TV was possible. Colour TV, like colour printing and colour photography (both demonstrated in the 1880s), relies on the fact that any colour seen in nature can be obtained by mixing three primary colours. For light, these colours are red, green and blue, and the primary colours used by painters are the paint colours (red, yellow and blue) that absorb these light colours. To obtain colour TV, then, you must display together three pictures, one consisting of red light, one of blue light and one of green light. This implies that the colour TV camera must generate three separate signals from the red, blue and green colours in an image.

All early colour TV systems worked on what is called a *sequential* system, meaning that the colours were neither transmitted nor seen at the same time. A typical method was the frame sequential system. Each picture frame was transmitted three times, using a different colour filter for each of a set of three views so that though what was transmitted was black and white,

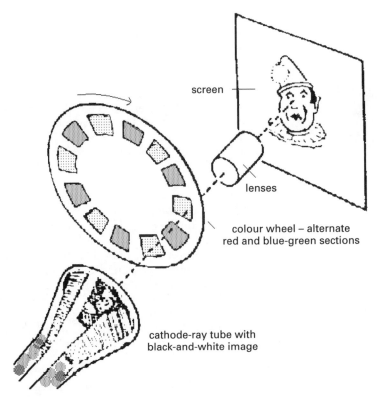

screen

lenses

colour wheel – alternate
red and blue-green sections

cathode-ray tube with
black-and-white image

Figure 8.10 A simple two-colour CRT-plus-wheel system, as demonstrated in 1942. Each picture is displayed for two frames, one in each colour, and the changes are rapid enough for the eye to see the average colour.

each of a set of three frames was different because it had been shot through a colour filter, one for each primary colour. At the receiver, a large wheel was spinning between the viewer and the TV screen, and this wheel carried a set of colour filters. The synchronisation was arranged so that the red filter would be over the CRT at the time when the frame containing the red image was being transmitted, so that this filter action put the colour into the transmitted monochrome picture. The main snag with this system is that the frames must be transmitted at a higher rate to avoid flickering, and there are also problems with compatibility and with the synchronisation of the wheel (and its size!). Figure 8.10 shows the principle of a two-colour system that was demonstrated by Baird in 1941. Baird later developed a CRT that could display two-colour mixtures simultaneously, but no one took up the idea (which worked) and Baird died shortly after the demonstrations.

Several early TV systems were devised to show still colour pictures, but the first commercially transmitted colour TV signals were transmitted in 1948 by CBS in New York, using a combination of electronic and mechanical methods. The system was not successful, and a commission on colour TV decided that no scheme could be licensed unless it was compatible. In other words, anyone with a monochrome receiver had to be able to see an acceptable picture in black and white when watching a colour broadcast, and anyone with a colour receiver had to be able to see an acceptable black and white picture when such a picture was being transmitted (to allow for the use of the huge stock of black-and-white films that TV studios had bought).

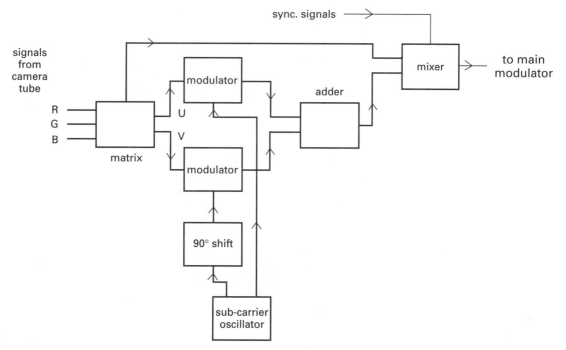

Figure 8.11 The PAL system. A block diagram, simplified, of the processes at the transmitter.

Colour TV, like colour photography, depends on detecting the red, green and blue light amplitudes in an image, since all natural colours can be obtained from mixtures of these pure colours. A colour TV camera must therefore produce three separate signals which can be R, G and B or mixtures from which separate R, G and B signals can be obtained. Though early TV systems had been able to produce pictures using sequential colour, these systems were not compatible with monochrome transmissions and were abandoned in favour of a simultaneous system in 1952.

RCA had been working on simultaneous systems throughout the 1940s, and its demonstrations in 1952 convinced the commission, the National Television Standards Committee (NTSC), that the RCA system was suitable. All other systems have taken this system as a basis, but differences in details mean that NTSC TV pictures are not compatible with the PAL colour system used in Germany, the UK and in other parts of Western Europe, or the French SECAM system used in France, former French colonies, and Eastern Europe. This is also why you cannot exchange videocassettes with friends overseas. Let's start with the portions that are common to all systems.

The image is split into red, green and blue components, using prisms, so that three separate camera tubes or semiconductor detectors can each produce a signal for one primary colour. This is the starting point for the block diagram of Figure 8.11. The signals are mixed to form three outputs. One is a normal monochrome signal, called the Y-signal. This is a signal that any monochrome receiver can use and it must be the main video signal so as to ensure compatibility. The other two outputs of this mixer are called *colour-*

difference signals, obtained by using different mixtures of the R, G, and B signals. These colour difference signals are designed to make the best use of the transmission system and are of lower amplitude than the monochrome signal, because in any TV signal, the monochrome portion always carries much more information. Colour is less important, and all the fine details of a picture can be in monochrome only because the eye isn't so sensitive to colour in small areas. These colour-difference signals are lettered as U and V.

The colour difference signals are transmitted using a sub-carrier, a method that we have looked at already in connection with stereo radio. This time, the use of the sub-carrier is much more difficult. To ensure compatibility, the sub-carrier frequency must be within the normal bandwidth of a monochrome signal, so that you would expect it to cause a pattern on the screen. As it happens, by choosing the sub-carrier frequency carefully, and keeping its amplitude low, it is possible to make the pattern almost invisible. The other problem is that this sub-carrier has to be modulated with *two* signals, and this is done by modulating one colour-difference signal on to the sub-carrier directly (using amplitude modulation) and then using a phase-shifted version of the sub-carrier, with phase shifted by 90°, and modulating the other colour difference signal on to this phase-shifted sub-carrier.

Modulating two waves of the same frequency but with 90° phase difference is equivalent to modulating both amplitude and phase of the same carrier.

If you remember that a 90° phase shift means that one wave is at zero when the other is at a maximum, you will appreciate that adding these two modulated sub-carriers together does not cause them to interfere with each other. The doubly-modulated sub-carrier can now be added to the monochrome signals (remember that the sub-carrier has a lower amplitude) and the synchronising signals added. This forms the colour composite video waveform which can be modulated on to the main carrier. What this amounts to is that the amplitude of the modulated signal represents the saturation of colour and the phase of the modulated signal represents the hue.

The hue is the colour, the wavelength of light, and saturation measures how intense the hue is. Most natural colours have low saturation values, meaning that these colours are heavily diluted with white.

Synchronising is not just a matter of the line and field pulses this time. A colour receiver has to be able to generate locally a copy of the sub-carrier in the correct phase, and to ensure this, some nine cycles of the subcarrier are transmitted along with each line synchronising pulse, using a time when the normal monochrome signal has a gap between the line synchronising pulse and the start of the line signal, Figure 8.12. This interval is called the *back porch* of the synchronising pulse. The illustration also shows how some coloured bars appear on a signal, as viewed by an oscilloscope (see Chapter 13). The sub-carrier in these signals is represented by shading. If the sub-carrier were not present, these bars would appear in shades of grey.

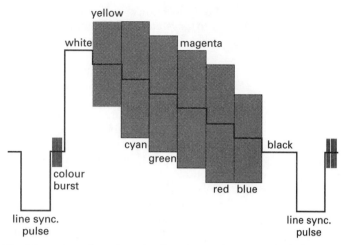

Figure 8.12 The colour burst and colour bars. The colour burst consists of ten cycles of sub-carrier located in the 'back porch' of each line sync. pulse. The diagram also shows the waveforms corresponding to colour bars – this cannot, of course, indicate the phase of the sub-carrier.

To ensure compatibility, the normal monochrome signal must be modulated on to the main carrier in the usual way. The colour difference signals are then transmitted by modulating them on to a sub-carrier with a frequency that is carefully chosen to cause the minimum of interference with the monochrome signal. One colour-difference signal is amplitude-modulated on to the sub-carrier directly, and the other is amplitude-modulated on to a sub-carrier of the same frequency but phase shifted by 90°. These signals are added to the monochrome signal, and nine cycles of the 0° phase sub-carrier signal are inserted following each line synchronising pulse so that the receiver can locally generate a sub-carrier in the correct phase.

The three systems
Compatible colour TV broadcasting started in the USA in 1952, using a scheme which was very much as has been outlined here (though the colour difference signals were not formed in the same way). The NTSC system always suffered from colour problems, and particularly in the early days, viewers complained that they constantly needed to adjust the colour controls that were fitted to receivers, hence the old joke that NTSC stood for 'never twice the same colour'. The problem was that during transmission of the signals, changes in the phase of the carrier caused by reflections had a serious effect on the sub-carrier, causing the colour information to alter. This is because the phase of the colour signal carries the hue information. Though development has reduced these problems, the system has remained virtually unaltered.

By contrast, the colour TV systems used in Europe were designed much later with an eye to the problems that had been experienced in the USA. Though the principles of transmitting a monochrome signal along with a sub-carrier for two colour signals has not changed, the way that the colour sub-carrier is used has been modified, and so also has the composition of the colour signals. There are two main European systems, PAL and SECAM. The PAL system was evolved by Dr Bruch at Telefunken, and the SECAM system by Henri de France and a consortium of French firms. SECAM is used in France, in former French colonies, and in Eastern Europe, but the PAL system has been more widely adopted. Only the USA and Japan retain the old NTSC system.

The PAL system uses the colour difference signals that we call U and V, with the colour mix carefully chosen so that these signals need only a small (and equal) bandwidth. What makes the essential difference, however, is that the V signal, which carries most of the hue information, is inverted on each even-numbered line, and the signal that is used in the receiver is always the average of one line and the preceding line. If there has been a phase shift in the sub-carrier caused by conditions between the transmitter and the receiver, then subtracting the V signals of one line from the V signals of the following line will have the effect of cancelling out the changes. Since the V signals carry the hue information, this eliminates the changes of colour that were such a problem with the original NTSC signals.

By contrast, the SECAM system works by using frequency modulation of the sub-carrier, using the U signal on one line and the V signal on the next. As for the PAL system, the information of two lines has to be gathered up to provide the signals for each one line.

The European colour TV systems have been designed to avoid the problems caused by phase changes of the colour signal. The PAL system does this by inverting the V signal on alternate lines and averaging the signals at the receiver to cancel out the effects of phase change. SECAM operates by transmitting U and V signals on alternate lines. In either case, the colour information in two successive lines is always averaged.

Colour TV tubes

A colour TV receiver depends very heavily on the cathode-ray tube, and the type that is universally used now is the colour-stripe type. This has replaced the colour-dot type which was used in 1952 and which continued in use until the later 1960s. These tubes allow simultaneous colour output, meaning that the colours of a picture are all being displayed together rather than in a sequence, so that colour TV tubes have to use three separate electron guns, one for each primary colour.

Figure 8.13 shows a magnified view of a portion of the screen of a typical tube. The glowing phosphors are arranged as thin stripes, using three different materials that glow each in a primary colour when struck by electrons. These stripes are narrow, and typically a receiver tube would use at least 900 stripes across the width of the screen. Tubes for computer monitors use a much larger number, and this is reflected in the price of a monitor as compared to a TV receiver of the same screen size.

The tube contains three separate electron guns, Figure 8.14. Between the screen and the electron guns, and close to the screen there is a metal mask consisting of narrow slits. These slits are lined up so that the electrons from one gun will strike only the phosphor stripes that glow red; the electrons from the second gun will strike only the stripes that glow green, and the electrons from the third gun will hit only the blue stripes. We can therefore call the electron guns red, green and blue because these are the colours that they will produce on the screen. If we can obtain a set of video signals for these guns, identical to the original red, green and blue signals from the camera tubes, we should achieve a perfect copy of the TV image in colour.

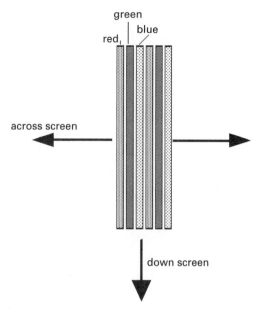

Figure 8.13 The arrangement of phosphor stripes on the screen of a colour display tube. Each dot viewed on the tube face consists of portions of a set of three stripes.

The display tube is the heart of any colour TV receiver. All modern colour CRTs use a set of phosphor stripes, arranged in a recurring R, G, B pattern. Close to this screen, a metal mask ensures that each of the three electron guns will project a beam that can hit only one colour of stripe each, so that signals to these guns will be R, G and B signals respectively.

The receiver circuits

A fair amount of the circuitry of a colour TV receiver is identical to that of a monochrome receiver. The superhet principle is used, and the differences start to appear only following the vision detector, where the video signal consists of the monochrome signal together with the modulated sub-carrier. The U and V colour-difference signals have to be recovered from this sub-carrier and combined with the monochrome signal so as to give three separate R, G and B signals that can be used on the separate guns of the colour tube.

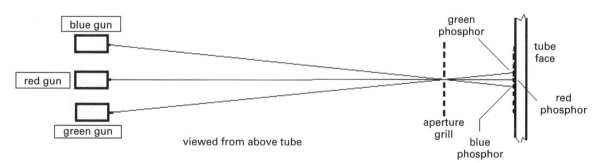

Figure 8.14 How the three guns of a colour tube are arranged. The aperture grill ensures that electrons from the 'red' gun strike only red phosphor stripes, and similarly for the other two guns.

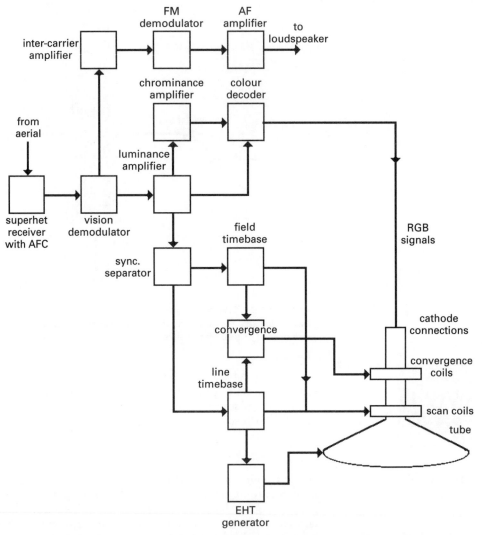

Figure 8.15 A general block diagram for a colour receiver. The convergence circuits keep the individual electron beams aimed at their respective phosphor stripes. Without convergence, the picture is satisfactory only at the middle of the screen.

A simplified outline of a receiver, showing the usual superhet stages as one block and neglecting differences between systems, is shown in Figure 8.15. The video signal has the 6 MHz sound signal filtered off, and the vision signal is amplified. At the output of this amplifier, other filters are used to separate the sub-carrier signal, which (in the UK) is at 4.43 MHz, with its sidebands, and the monochrome or *luminance* signal, with the sub-carrier frequencies greatly reduced by filtering, is separated off and amplified. The sub-carrier (*chrominance*) signal has to be demodulated in two separate circuits, because we have to recover two separate signals, U and V, from it. This requires a circuit in which the unmodulated sub-carrier frequency, in the correct phase, is mixed with the modulated sub-carrier signal. This type of circuit is called a *synchronous demodulator*.

A pulse is taken from the line timebase and used to switch on gates, circuits which will pass signal only for a specified interval. One of these

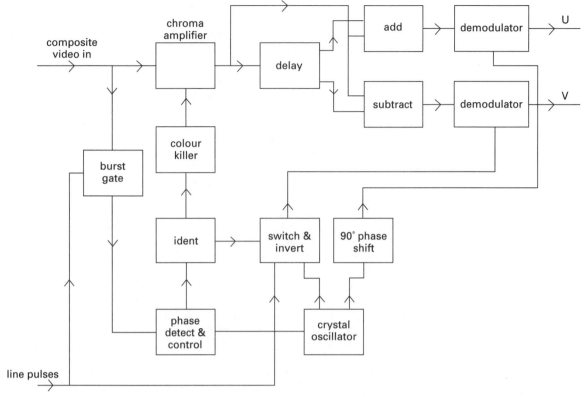

Figure 8.16 The decoding of the PAL type of signal. This is a very complicated process compared with the older NTSC system, but the results have borne out the wisdom of adopting the system for the UK.

gates allows the colour synchronising 'burst' to pass to the local sub-carrier oscillator to maintain it at the correct frequency and phase. This oscillator output is fed to a demodulator where the other input is the modulated sub-carrier, and this has the effect of recovering the V signal. The oscillator output is also phase shifted through 90° and this shifted wave is used in another demodulator to recover the U signal. The U, V and Y (mono-chrome) signals are then mixed to get the R, G and B signals that are fed to the CRT guns.

That's the simplest possible outline, and it's quite close to the original NTSC system for receivers. The PAL receiver incorporates more complications because at the camera end the V signal was inverted on each line, and the receiver needs to be able to combine the V signals from each pair of lines by storing the information of each line and combining it with the following line.

This is done using the block diagram of Figure 8.16, which shows part of the PAL receiver concerned with colour decoding. One output from the crystal oscillator is passed to an inverter circuit, and this in turn is control-led by a bistable switch. The word bistable in this context, means that the switch will flip over each time it gets a pulse input, and its pulse inputs are from the line timebase of the receiver, so that the switch is operated on each line. The action of the switch on the inverter will ensure that the oscillator signal is in phase on one line and will reverse (180° phase shift) on the next line and so on. Also shown in this diagram is an arrangement for an identification (*ident*) signal, taken from the burst signal, that makes certain that the bistable switch is itself operating in phase, and not inverting a signal that ought to be unchanged.

The sub-carrier signals are fed to a time delay. This is arranged to delay the signal by exactly the time of one line, and the conventional method is to use a glass block. The electrical signals are converted into ultrasonic sound-wave signals, and they travel through the glass to a pick-up where they are converted back into electrical signals. The delay is the time that these ultrasound signals take to travel through the glass block, and the dimensions of the glass are adjusted so that this is exactly the time of a line. Modern receivers can use electronic digital delay circuits, but these are outside the scope of this book. The delayed signal (the sub-carrier for the previous line) is added to the input signal (the sub-carrier for the current line) in one circuit and subtracted in a second circuit so as to produce averaged signals for the U and V demodulators that will contain no phase errors. Note that averaging by itself would not correct phase changes; it is the combination of averaging with the phase reversal of the V signals that ensures the correction.

In addition, there are embellishments on the circuits, and one of these is the *colour killer*. When the transmitted signals are in monochrome, any signal at the sub-carrier frequency would produce colour effects at the receiver, so that these circuits must be switched off for transmissions (old films, for example) that contain no colour. This is done by using the colour burst signals, so that when a colour burst is present, a steady voltage from a demodulator will ensure that the colour circuits are biased on and working. In the absence of the burst signal, the colour circuits are biased off, ensuring that any frequencies around 4.33 MHz (the colour sub-carrier frequency for the UK) in the picture signal do not cause colours to appear.

The colour processing for a receiver consists of separating off the signals at the sub-carrier frequency. A copy of the sub-carrier is generated and its phase and frequency corrected by the burst signal. This sub-carrier can be used in demodulators to recover the original U and V colour difference signals, and these can be combined with the luminance signal to form separate R, G and B colour signals.

For the PAL system, the sub-carrier used for the V signal must be inverted on each other line, and an ident signal is used to ensure that the inversions are in step. In addition, the colour signals are passed through a time delay of exactly the time of one line, so that the colour signals of one line can be combined with those of the previous line (with the V signal inverted) to provide averaged U and V signals.

Radar

Radar principles are so similar in many ways to those used for TV that the two can be dealt with here in a single chapter. Radar depends on the fact that the speed of electromagnetic waves in air is constant, identical to the speed of light at 300 metres per microsecond. If radio waves are sent out as a beam and are reflected from a target, the reflected wave will be detected some time after the transmitted wave, and the time delay will be equal to the time for the waves to travel to the target and back again. For example, a target at 300 metres distance will correspond to a delay time of 2 µs, and a target at 30 km distance will cause a return beam (an echo) to appear 200 µs later.

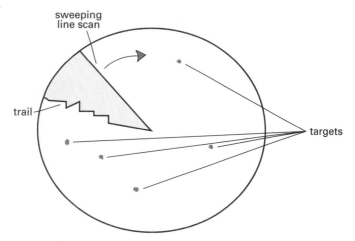

Figure 8.17 The appearance of a PPI display, showing the rotating scan. The tube is always of a long-persistence type, so that the scan line leaves a trail, and targets appear as slowly-fading blobs. The range is indicated by the distance from the centre of the screen, and the bearing is indicated by the angle of the target blobs.

The classical type of radar that has been used for more than 50 years is the *plan-position indicator* (PPI) which produces on a CRT a pattern in which the base station is represented by the centre of the screen, and the targets appear as bright spots. The distance of a target dot from the centre of the screen is proportional to the distance of the target from the transmitter, and the bearing of the target is given by the angle from a zero degree line drawn on the screen (or on a cursor that fits over the screen), Figure 8.17. Though there have been considerable developments in radar systems over the last 50 years, these basic principles remain almost unchanged.

The requirements for radar are the transmission of short pulses of very high frequency waves, the demodulation and amplification of the return signal, and the use of the return signal to display on a CRT. Though lower frequencies were used initially, the frequencies that are used are nowadays several GHz, and they are generated by vacuum devices called *magnetrons*. The magnetron is a form of oscillator which causes radio wave oscillations in much the same way as a flute player causes sound oscillations. The magnetron uses a magnet to make an electron beam take a circular path, and by steering this path across circular cavities in a block of metal, the beam can be made to oscillate at a frequency that is determined by the dimensions of the cavity. The cavity will typically be of a few millimetres radius, so that the waves that are generated have a wavelength of a few millimetres, corresponding to frequencies of several GHz. These frequencies are called the microwave range, and the familiar microwave oven uses a magnetron working at 2.45 GHz, which is a frequency that matches the natural frequency of vibration of molecules of water.

Radar makes use of waves in the GHz frequency range. These have a short wavelength, behave like light waves, and reflect from dense objects. Specialised vacuum tubes called magnetrons can generate these waves, usually in short bursts. When a burst of microwaves is beamed from an aerial and hits a target, the echo will return in a time that depends on the distance of the target.

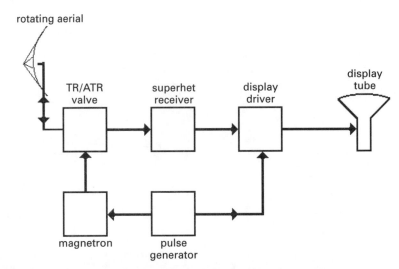

Figure 8.18 A block diagram for a PPI type of radar system.

The classic type of magnetron is pulsed by switching the power on and off, typically for one microsecond or less in each millisecond. Since the magnetron is switched on for only one thousandth of the time when it is being used, the amount of power in each pulse can be very large without causing cooling problems. For example, if the power in a pulse is 1 MW (one million watts), the average power might be just 1 kW (1000 watts), small enough to cope by air or water cooling of the magnetron.

Figure 8.18 illustrates a block diagram for a PPI radar station. The clock circuits generate the pulses, setting the time between pulses (this is more usually expressed as a frequency, the pulse repetition rate or PRF). One output from the clock is used to generate high-voltage pulses of the duration that is required, and these pulses are applied to the magnetron by way of a modulator. The pulse of microwaves passes through the TR/ATR valve, which separates the transmitting circuits from the receiving circuits. This is an essential part of any radar in which a single aerial is used for both transmission and reception, because with typical transmitter pulse power of 1 MW and typical received signal of a few µW, the receiver would be vaporised if the transmitter signal reached it.

The aerial consists of a dish similar to the type that is so familiar for satellite reception (which uses around 11 GHz frequencies). Radar dishes are usually larger because space is not usually a problem, and frequencies lower than 11 GHz are being used. In addition, the dish rotates so that it aims its microwave beam in a slightly different direction for each pulse. When the pulse of microwaves has been transmitted, the TR/ATR valves close off the transmitter circuits from the aerial and opens the receiver circuits. The returning echo is received on the dish and its signal is passed to the receiver which, as usual, is a superhet design. At the demodulator, the signal consists of a pulse identical in width to the original pulse. This signal is then gated by another pulse that is derived from the clock circuit. The purpose of this is to keep unwanted signals from the display by allowing signals to pass only for a specified time after sending out a pulse.

The display consists of the CRT with a scan that is radial, deflecting the beam from the centre of the tube out to the edges. At one time, the scanning coils were rotated in step with the rotation of the aerial dish, but by the

1950s it was more usual to keep the scanning coils steady and alter the phasing of the scanning signals so that the direction of deflection of the beam rotates in step with the rotating aerial dish. The received pulse is used to brighten the beam, providing a target display, and a graticule (a transparent scale) can be used to read direction and distance. On modern radar displays, computer techniques place a reference number against each target so as to make it easier to follow the movement of a target, and figures for the size, distance, bearing and speed of a target can be obtained from the computer system.

Doppler radar

Doppler radar is used for height and speed measurement, and it depends on a continuous microwave output rather than the use of pulses. On a Doppler system, microwaves are sent out and received continuously, and the two waves are mixed together (using circuits that ensure that both are of the same amplitude). For a target that is not moving, the returning waves have exactly the same frequency as the transmitted waves, but when a target is moving the returning wave has a slightly different frequency. This returning frequency is higher if the target is moving towards the aerial, and is lower if the target is moving away from the aerial. The difference between the transmitted and the returned frequency can be used to measure the speed of the target to or from the aerial, and on modern equipment, a computer is used to work out speed. For aircraft height measurements, the aerial points vertically downwards and the phase of the signal returned from the ground is compared with the phase of the transmitted signal, and the difference used to find the height. This is much more precise than the barometer type of altimeters that were in use into the 1950s even on military aircraft.

The PPI type of radar uses a rotating aerial and a display that consists of a scan line rotating in synchronism. The returning echo pulse is amplified and use to brighten the trace, so that the distance from the centre of the CRT represents distance and the bearing of the trace is the same as the bearing of the target. Doppler radar uses continuous rather than pulsed waves and measures the speed of a target from the change of frequency of the echo signal. Phase changes can be used to measure distance, notably small distances such as are required for height measurements.

Chapter 9   Digital signals

Voltage levels

A digital signal is one in which a change of voltage, and the time at which it occurs, are of very much more importance than the precise size of the change or the exact shape of the waveform. All of the waveforms used in digital circuits are steep-sided pulses or square waves and it is the change of voltage that is significant, not the values of voltage. For that reason, the voltages of digital signals are not referred to directly, only as 1 and 0. The important feature of a digital signal is that each change is between just two voltage levels, typically 0 V and +5 V, and that these levels need not be precise. In this example, the 1 level can be anything from 2.4 V to 5.2 V, and the 0 level anything from 0 V to 0.8 V. By using 0 and 1 in place of the actual voltages, we make it clear that digital electronics is about numbers, not waveforms.

The importance of using just two digits is that this is ideally suited to electronic devices. A transistor, whether a bipolar or FET type, can be switched either fully on or fully off, and these two states can be ensured easily, much more easily than any other states that depend on a precise bias. By using just these two states, then, we can avoid the kind of errors that would arise if we tried to make a transistor operate with, say, ten levels of voltage between two voltage extremes. By using only two levels, the possibility of mistakes is made very much less. The only snag is that any counting that we do has to be in terms of only two digits, 0 and 1, and counting is the action that is most needed in digital circuitry.

Counting with only two digits means using a scale of two called the *binary scale*, in place of our usual scale of ten (a denary scale). There is nothing particularly difficult about this, because numbers in this scale are written in the same way as ordinary (denary) numbers. As with denary numbers, the position of a digit in a number is important. For example, the denary number 362 means three hundreds, six tens and two units. The position of a digit in this scale represents a power of 10, with the right-hand position (or *least-significant* position) for units, the next for tens, the next for hundreds (ten squared), then thousands (ten cubed) and so on.

For a scale of two, the same scheme is followed. In this case, however, the positions are for powers of two, as units, twos, fours (two squared), eights (two cubed) and so on. Table 9.1 shows powers of two as place numbers and their denary equivalents, and the text shows how a binary number can be converted to denary form and a denary number to binary form.

Converting binary to denary: For each 1 in the number, write down the denary number for that place, then add.

Example: Binary 10001110110. This uses place number 0 to 10, with 1s in positions 1, 2, 4, 5, 6 and 10. This corresponds to denary numbers 1024 + 64 + 32 + 16 + 4 + 2 = 1142

Table 9.1 Binary and denary numbers

Place no.	Denary	Place no.	Denary	Place no.	Denary	Place no.	Denary
0	1	8	256	16	65 536	24	16 777 216
1	2	9	512	17	131 072	25	33 554 432
2	4	10	1 024	18	262 144	26	67 108 864
3	8	11	2 048	19	524 288	27	134 217 728
4	16	12	4 096	20	1 048 576	28	268 435 456
5	32	13	8 192	21	2 097 152	29	536 870 912
6	64	14	16 384	22	4 194 304	30	1 073 741 824
7	128	15	32 768	23	8 388 608	31	2 147 483 648

Converting denary to binary: Divide the number by two, and write the remainder (which must be 0 or 1) at the side. Now divide the last result also by two, again writing the remainder at the side. Repeat until the last remainder (which is always 1) has been found, then read the numbers from *bottom to top*.

Example: Denary 875

$875 \div 2 = 437$ remainder 1
$437 \div 2 = 218$ remainder 1
$218 \div 2 = 109$ remainder 0
$109 \div 2 = 54$ remainder 1
$54 \div 2 = 27$ remainder 0
$27 \div 2 = 13$ remainder 1
$13 \div 2 = 6$ remainder 1
$6 \div 2 = 3$ remainder 0
$3 \div 2 = 1$ remainder 1
$1 \div 2 = 0$ remainder 1

Binary number is 1101101011

In a binary number such as 1100, the last zero is the *least-significant bit*, and the first 1 is the *most significant bit*. Most arithmetic actions start with the least significant bit and work towards the most significant bit shifting left through the digits.

Digital circuits are switching circuits, and the important feature is fast switching between the two possible voltage levels. Most digital circuits would require a huge number of transistors to construct in discrete form, so that digital circuits make use of ICs, mostly of the MOSFET type. The use of integrated construction brings two particular advantages to digital circuits. One is that circuits can be very much more reliable than when separate components are used (in what are called *discrete* circuits). The other advantage is that very much more complex circuits, with a large number of components can be made as easily in integrated form (once the master templates have been made) as simple circuits.

Because digital systems are based on counting with a scale of two, their first obvious applications were to calculators and computers, topics we shall deal with in Chapter 12. What is much less obvious is that digital signals can be used to replace the analogue type of signals that we have become accustomed to. This is the point we shall pay particular attention to in this chapter, because some of the most startling achievements of digital circuits are where digital methods have replaced analogue methods, such as the audio CD, the video disc and in the digital TV systems that will eventually replace the methods that we have grown up with.

As it happens, the development of digital ICs has had a longer history than that of analogue IC devices. When ICs could first be produced, the manufacturing of analogue devices was extremely difficult because of the difficulty of ensuring correct bias and the problems of power dissipation. Digital IC circuits, using transistors that were either fully off or fully on, presented no bias problems and had much lower dissipations. In addition, circuits were soon developed that reduced dissipation still further by eliminating the need for resistors on the chip. Digital ICs therefore had a head start as far as design and production was concerned, and because they were immediately put into use, the development of new versions of digital ICs was well under way before analogue ICs made any sort of impact on the market.

Digital signals consist of rapid transitions between two voltage levels that are labelled as 0 and 1 – the actual values are not important. This form of signal is well suited to active components, because the 0 and 1 voltages can correspond to full-on and cut-off conditions respectively, each of which causes very little dissipation. ICs are ideally suited to digital signal use because complex circuits can be manufactured in one set of processes, dissipation is low, reliability is very high and costs can be low.

Digital recording

Given the advantages of digital signals as far as the use of transistors and ICs is concerned, what are the advantages for the processing of signals? The most obvious advantage relates to tape or any other magnetic recording. Instead of expecting the magnetisation of the tape to reproduce the varying voltage of an analogue signal, the tape magnetisation will be either maximum in one direction or maximum in the other. This is a technique called 'saturation recording' for which the characteristics of most magnetic recording materials are ideally suited. The precise amount of magnetisation is no longer important, only its direction. This, incidentally, makes it possible to design recording and replay heads rather differently so that a greater number of signals can be packed into a given length of track on the tape. Since the precise amount of magnetisation is not important, linearity problems disappear.

Noise problems are also eliminated. Tape noise consists of signals that are, compared to the digital signals, far too small to register, so that they have no effect at all on the digitally recorded signals. This also makes tapes easier to copy, because there is no degradation of the signals caused by copying noise, as there always is when conventional analogue recorded tapes are copied. Since linearity and noise are the two main problems of any tape (or other magnetic) recording system it is hardly surprising that recording studios have rushed to change over to digital tape mastering. The surprising thing is that it has been so late in arriving on the domestic scene, because the technology has been around for long enough, certainly as long as that of videotape recording. A few (mainly Betamax) video recorders provided for making good-quality audio recordings of up to eight hours on videotape, but this excellent facility was not taken up by many manufacturers, and died out when VHS started to dominate the video market in the UK. It has reappeared on some NICAM machines more recently.

The advantages that apply to digital recording with tape apply even more forcefully to discs. The accepted standard method of placing a digital signal on to a flat plastic disc is to record each digit bit as a tiny pit on the otherwise flat surface of the disc, and interpret a digital 1 as the change of reflection of a laser beam. Once again, the exact size of the pit is unimportant as long as it can be read by the beam, and only the number of pits is used to carry signals. We shall see later that the process is by no means so simple as this would indicate, and the CD is a much more complicated and elaborate system than the tape system (DAT) that is now available, though at prohibitive prices, in the UK. The basic principles, however, are simple enough, and they make the system immune from the problems of the LP disc. There is no mechanical cutter, because the pits have been produced by a laser beam which has no mass to shift and is simply switched on and off by the digital signals. At the replay end of the process, another (lower-power) laser beam will read the pattern of pits and once again this is a process which does not require any mechanical movement of a stylus or any pick-up mechanism, and no contact with the disc itself.

As with magnetic systems, there is no problem of linearity, because it is only the number of pits rather than their shape and size that counts. Noise exists only in the form of a miscount of the pits or as confusion over the least significant bit of a number, and as we shall see there are methods that can reduce this to a negligible amount. Copying of a disc is not so easy as the moulding process for LPs, and mass production requires enormous capital investment and costly inspection processes, all of which are threatened by the advent of digital tape systems. A CD copy, however, is much less easily damaged than its LP counterpart, and even discs that look as if they are badly damaged will play with no noticeable effects on the quality of the sound – though such a disc will sometimes skip a track. The relative immunity to wear is a very strong point, because this allows CDs to recoup their high initial manufacturing costs when they are used in juke-box applications or for the inescapable deluge of recorded sound in restaurants and other public places. This has also allowed CD libraries to flourish, because each disc can earn its keep in loan fees and can afterwards be sold at a good price because it can still be played with no noticeable deterioration in quality. CD recording machines, mainly intended for computer CD-ROM, are now no more expensive than a good computer system.

Summary

Outside computing, digital systems are best known for the CD and to a lesser extent in magnetic recording. Virtually every recording now made uses digital tape systems for mastering, however, so that digital methods are likely to be used at each stage in the production of a CD (this is marked by a DDD symbol on a CD). Magnetic recording uses a saturation system, using one peak of magnetisation to represent 1 and the opposite to represent 0. Disc recording uses tiny pits on a flat surface to represent a 1 and the flat surface to represent 0.

Conversions

No advantages are ever obtained without paying some sort of price, and the price to be paid for the advantages of digital recording, processing and reproduction of signals consists of the problems of converting between analogue and digital signal systems, and the increased rate of processing of data. For example, a sound wave is not a digital signal, so that its electrical counterpart must be converted into digital form. This must be done at some stage where the electrical signal is of reasonable amplitude, several volts, so that

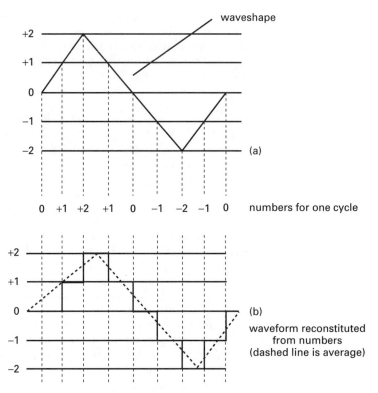

Figure 9.1 Quantising a waveform. The point where the waveform changes to a new level is represented by the number of that level, so that the waveform is coded as a stream of numbers. Reversing the process produces a shape that when smoothed (averaged) provides a recognisable copy of the input even for this very crude five-level system.

any noise that is caused will be negligible in comparison to the signal amplitude. That in itself is no great problem, but the nature of the conversion is.

What we have to do is to represent each part of the wave by a number where the value is proportional to the voltage of the waveform at that point. This involves us right away into the two main problems of digital systems, resolution and response time. Since the conversion to and from sound waves is the most difficult challenge for digital systems, we shall concentrate on it here. By comparison, radar and even television are systems that were almost digital in form even from the start. For example, the TV line waveform consists of the electrical voltage generated from a set of samples of brightness of a line of a scanned image.

To see just how much of a problem the conversion of sound waves is, imagine a system that used only the numbers –2 to +2, on a signal of 4 V total peak-to-peak amplitude. If this were used to code a waveform (shown as a triangular wave for simplicity) as in Figure 9.1(a) then since no half-digits can exist, any level between –0.5 V and +0.5 V would be coded as 0, any signal between +0.5 V and +1.5 V as 1 and so on, using ordinary denary numbers rather than binary numbers to make the principle clearer. In other words, each part of the wave is represented by an integer (whole) number between –2 and +2, and if we plotted these numbers on the same graph scale then the result would look as in Figure 9.1(b). This is a 'block' shape of wave, but recognisably a wave which if heavily smoothed would be something like the original one. We could say that this is a five-level quantisation of the wave, meaning that the infinite number of voltage levels of the original wave have been reduced to just five separate levels. This is a very crude quantisation, and the shape of a wave that has been quantised to a larger number of levels is a much better approximation to the original. The larger the number of levels, the closer the wave comes to its original pattern, though we are cheating in a sense by using a sinewave as an illustration, since this is the simplest type of wave to convert in each direction. Nevertheless, it is clear that the greater the number of levels that can be expressed as different numbers then the better is the fidelity of the sample.

Quantisation means the sampling of a waveform so that the amplitude of each sample can be represented by a number.

In case you feel that all this is a gross distortion of a wave, consider what happens when an audio wave of 10 kHz is transmitted by medium-wave radio, using a carrier wave of 500 kHz. One audio wave will occupy the time of 50 radio waves, which means in effect that the shape of the audio wave is represented by the amplitudes of the peaks of 50 radio waves, a 50-level quantisation. You might also like to consider what sort of quantisation is involved when an analogue tape system uses a bias frequency of only 110 kHz, as many do. The idea of carrying an audio wave by making use of samples is not in any way new, and is inherent in amplitude-modulation radio systems which were considered reasonably good for many years. It is equally inherent in frequency modulation, and it is only the use of a fairly large amount of frequency change (the peak deviation) that avoids this type of quantisation becoming too crude. Of all the quantised ways of carrying an audio signal, in fact, FM is probably the most satisfactory, and FM methods are often adopted for digital recording, using one frequency to represent a 0 and another to represent a 1.

The conversion of a waveform into a set of digital signals starts with quantisation of the wave to produce a set of numbers. The greater the number of quantisation levels, the more precise is the digital representation, but excessive quantisation is wasteful in terms of the time required.

This brings us to the second problem, however. Because the conversion of an audio wave into a set of digits involves sampling the voltage of the wave at a large number of intervals, the digital signal consists of a large set of numbers. Suppose that the highest frequency of audio signal is sampled four times per cycle. This would mean that the highest audio frequency of 20 kHz would require a sampling rate of 80 kHz. This is not exactly an easy frequency to record even if it were in the form of a sinewave, and the whole point of digital waveforms is that they are not sinewaves but steep-sided pulses which are considerably more difficult to record. From this alone, it is not difficult to see that digital recording of sound must involve rather more than analogue recording.

The next point is the form of the numbers. We have seen already that numbers are used in binary form in order to allow for the use of only the two values of 0 and 1. The binary code that has been illustrated in this chapter is called 8–4–2–1 binary, because the position of a digit represents the powers of two that follow this type of sequence. There are, however, other ways of representing numbers in terms of 0 and 1, and the main advantage of the 8–4–2–1 system is that both coding and decoding are relatively simple. Whatever method is used, however, we cannot get away from the size of a binary number. It is generally agreed that modern digital audio should use a 16-bit number to represent each wave amplitude, so that the wave amplitude can be any of up to 65536 values. For each sample that we take of a wave, then, we have to record 16 digital signals, each 0 or 1, and all 16 *bits* will be needed in order to reconstitute the original wave.

A bit is short for a binary digit, a 0 or 1 signal.

This is the point on which so many attempts to achieve digital coding of audio have foundered in the past. As so often happens, the problems could be more easily solved using tape methods, because it would be quite feasible to make a 16-track tape recorder using wide tape and to use each channel for one particular bit in a number. This is, in fact, the method that can be used for digital mastering where tape size is not a problem, but the disadvantage here is that for original recordings, some 16 to 32 separate music tracks will be needed. If each of these were to consist of 16 digital tracks the recorder would, to put it mildly, be rather overloaded. Since there is no possibility of creating or using a 16-track disc, the attractively simple idea of using one track per digital bit has to be put aside. The alternative is serial transmission and recording.

Serial means one after another. For 16 bits of a binary number, serial transmission means that the bits are transmitted in a stream of 16 separate signals of 0 or 1, rather in the form of separate signals on 16 channels at once. Now if the signals are samples taken at the rate of 60 kHz, and each signal requires 16 bits to be sent out, then the rate of sending digital signals is 16 × 60 kHz, which is 960 kHz, well beyond the rates for which ordinary tape or disc systems can cope. As it happens, we can get away with slower sampling rates, as we shall see, but this doesn't offer much relief because there are further problems. When a parallel system is used, with one channel for each bit, there is no problem of identifying a number, because the bits are present at the same time on the 16 different channels. When bits are sent one at a time, though, how do you know which bits belong to which number? Can you be sure that a bit is the last bit of one number or is it the first bit of the next number? The point is very important because when the 8–4–2–1 system is used, a 1 as the most important bit means a value of 32768, but a 1 as the least important bit means just 1. The difference in terms of signal amplitudes is enormous, which is why codes other than the 8–4–2–1 type are used industrially. The 8–4–2–1 code is used mainly in computing because of the ease with which arithmetical operations can be carried out on numbers that use this code.

Even if we assume that the groups of 16 bits can be counted out perfectly, what happens if one bit is missed or mistaken? At a frequency of a megahertz or so it would be hopelessly optimistic to assume that a bit might not be lost or changed. There are tape dropouts and dropins to consider, and discs cannot have perfect surfaces. At such a density of data, faults are inevitable, and some methods must be used to ensure that the groups of 16 bits, called 'words', remain correctly gathered together. Whatever method is used must not compromise the rate at which the numbers are transmitted, however, because this is the sampling rate and it must remain fixed. Fortun-

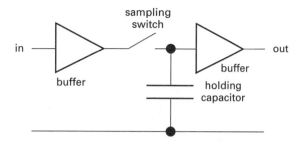

Figure 9.2 A sample and hold circuit, using a switch to represent the switching actions that would be carried out using MOSFETs. When the switch opens, the voltage on the capacitor is a sampled waveform voltage for that instant, and the conversion to digital form of this voltage at the output of the sample and hold circuit must take place before the next sample is taken.

ately, the problems are not new nor unique to audio; they have existed for a long time and been tackled by the designers of computer systems. A look at how these problems are tackled in simple computer systems gives a few clues as to how the designers of audio digital system went about their task.

The problem of conversion from analogue to digital signals for sound waves is the rate of conversion that is needed. The accepted number of digits per sample is 16, and a well-known rule (Shannon's law) states that the sampling rate must be at least twice the highest component of the analogue signal. For a maximum of 20 kHz, this means a sampling rate of 40 kHz, and for 16-bit signals, this requires a rate of 16×20 thousand bits per second, which is 320 000 bits. These bit signals have to be recorded and transmitted in serial form, meaning one after another.

Analogue to digital

Converting from an analogue into a digital signal involves the quantisation steps that have been explained above, but the mechanism needs some explanation in the form of block diagrams. All analogue to digital conversions start with a sample and hold circuit, illustrated in block form in Figure 9.2. The input to this circuit is the waveform that is to be converted to digital form, and this is taken through a buffer stage to a switch and a capacitor. While the switch is closed, the voltage across the capacitor will be the waveform voltage – the buffer ensures that the capacitor can be charged and discharged without taking power from the signal source. At the instant when the switch opens, the voltage across the capacitor is the waveform voltage, and this will remain stored until the capacitor discharges. Since the next stage is another buffer, it is easy to ensure that the amount of discharge is negligible. While the switch is open and the voltage is stored, the conversion of this voltage into digital form (quantisation) can take place, and this action must be completed before the switch closes again at the end of the sampling period.

In this diagram, a simple mechanical switch has been shown but in practice this switch action would be carried out using MOSFETs which are part of the conversion IC. To put some figures on the process, at a sampling rate of 44.1 kHz as is used for CDs, the hold period cannot be longer than 22 µs, which looks long enough for conversion – but read on!

The conversion can use a circuit such as is outlined in Figure 9.3, and which very closely resembles the diagram of a digital voltmeter illustrated

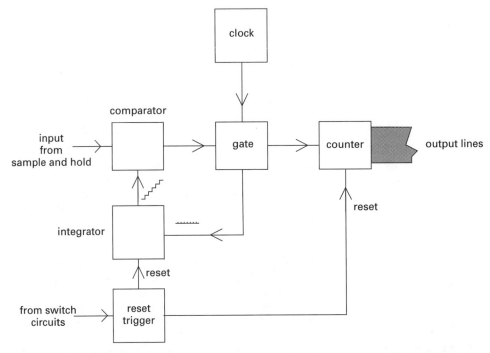

Figure 9.3 A typical analogue to digital (A/D) converter. The gate passes clock pulses that are integrated until the output of the integrator equals the comparator input voltage. The pulse count is the output of the counter, the digital signal. The circuit then resets for the next count.

in Chapter 11. The clock pulses are at a frequency that is much higher than the sampling pulses, and while a voltage is being held at the input, the clock pulses pass through the gate and are counted. The clock pulses are also the input to the integrator, whose output is a rising voltage. When the output voltage from the integrator reaches the same level (or slightly above the level) as the input voltage, the gate shuts off, and at this point the count is used as the digital signal. The reset pulse (from the sample-and-hold switch circuit) then resets the counter and the integrator so that a new count can start for the next sampled voltage. The clock pulse must be at a rate that will permit a full set of pulses to be counted in a sampling interval. For example, if the counter uses 8-bit output, corresponding to a count of 65538 (which is 2^8), and the sampling time is 20 μs then it must be possible to count 65536 clock pulses in 20 μs, giving a clock rate of 3.27 GHz. This is not a rate that would be easy to supply or to work with, so that the conversion process is not quite as simple as has been suggested here. For CD recording, more advanced A/D conversion methods are used, such as the successive approximation method, in which the input voltage is first compared to the voltage corresponding to the most significant digit, then successively to all the others, so that only eight comparisons are needed for an 8-bit output rather than 65536. If you are curious about these methods, see the book *Introducing Digital Audio* from PC Publishing.

The sampling rate for many analogue signals can be much lower than the 44.1 kHz that is used for CD recording, and most industrial processes can use much lower rates. ICs for these A/D conversions are therefore mostly of the simpler type, taking a few milliseconds for each conversion. Very

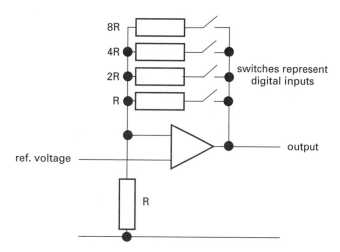

Figure 9.4 A simple voltage-adding D/A converter.

high-speed ('flash') converters are also obtainable that can work at sampling rates of several MHz.

Digital to analogue

Converting digital to analogue is easy enough if a small number of bits are being used and if the rate of conversion is low. Figure 9.4 shows a simple method that uses an operational amplifier for a 4-bit digital signal. If we imagine that the switches represent a set of digital inputs (closed = 1, open = 0) then the digital signals will close feedback paths for the operational amplifier, so that the output voltage will be some multiple of the input (reference) voltage that is set by the total resistance in the feedback path. In practice, of course, the switches are MOSFETs which are switched on or off by the digital inputs on four lines in this example.

The snag with this system, which is quite adequate for 4-bit conversions is that it presents impossible problems when you try to use it for eight or more bits. The snag is that the resistor values become difficult to achieve, and the tolerance of resistance is so tight that manufacturing is expensive. Other methods have therefore been devised, such as current addition and bitstream converters, both beyond the scope of this book. Converters for 16 bits are now readily available at reasonable prices.

Conversion between analogue and digital signals is straightforward if sampling rates are low and the number of bits is small. More elaborate methods are needed for fast conversion and for 16-bit operation, but the problems have been solved by the IC manufacturers and a large range of conversion ICs can be bought off the shelf.

Serial transmission

To start with, when computers transmit data serially, the 'word' that is transmitted is not just the group of digits that is used for coding a number. For historic reasons, computers transmit in units of eight bits, called a byte, rather than in 16-bit words, but the principles are equally valid. When a byte is transmitted over a serial link, using what is called asynchronous methods, it is preceded by one 'start bit' and followed by one or two (according to the system that is used) 'stop bits'. Since the use of two stop

Signal byte	00110111	01010010	00101110	01101001
Even parity added	1	1	0	0
Signal sent	10110111	11010010	00101110	01101001
Received byte	10100111	11010010	00101110	01101001
Parity check	odd	even	odd	even
Result	error	OK	error	OK

Figure 9.5 Using parity to check that a byte has been received correctly. Each number uses seven bits, and one extra parity bit is added, in this case to make the count of 1s an even number. On reception, the byte can be checked to find if its parity is even.

bits is very common, we will stick to the example of one start bit, eight-number bits and two-stop bits. The start bit is a 0 and the stop bits are 1s, so that each group of 11 bits that are sent will start with a 0 and end with two 1s. The receiving circuits will place each group of 11 bits into a temporary store and check for these start and stop bits being correct. If they are not, then the digits as they come in are shifted along until the pattern becomes correct. This means that an incorrect bit will cause loss of data, because it may need several attempts to find that the pattern fits again, but it will not result in every byte that follows being incorrect, as would happen if the start and stop bits were not used.

The use of start-and-stop bits is one method of checking the accuracy of digital transmissions, and it is remarkably successful, but it is just one of a number of methods. In conjunction with the use of start-and-stop bits, many computer systems also use what is known as parity, a method of detecting one-bit errors. In a group of eight bits, only seven are normally used to carry data and the eighth is spare. This redundant bit is made to carry a checking signal, which is of a very simple type. We will illustrate how it works with an example of what is termed even parity. Even parity means that the number of 1s in a group of eight shall always be even. If the number is odd, then there has been an error in transmission and a computer system may be able to make the transmitting equipment try again. When each byte is sent, the number of 1s is counted. If this number is even, then the redundant bit is left as a 0, but if the number is odd, then the redundant bit is made a 1, so that the group of eight now contains an even number of 1s. At the receiver, all that is normally done is to check for the number of 1s being even, and no attempt is made to find which bit is at fault if an error is detected. The redundant bit is not used for any purpose other than making the total number even. The process is illustrated in Figure 9.5.

Parity, used in this way, is a very simple system indeed, and if two bits in a byte are in error it is possible that the parity could be correct though the transmitted data was not. In addition, the parity bit itself might be the one that was affected by the error so that the data is signalled as being faulty even though it is perfect. Nevertheless parity, like the use of start bits and stop bits, works remarkably well and allows large masses of computer data to be transmitted over serial lines at reasonably fast rates. What is a reasonably fast rate for a computer is not, however, brilliant for audio, and even for the less-demanding types of computing purposes, the use of parity is not really good enough, and much better methods have been devised. The rates of sending bits serially for computing purposes range from the abysmally slow 110 bits per second to the reasonably fast 19 600 bits per second. Even this fast rate is very slow by the standards that we have been talking about, so it is obvious that something rather better is needed for audio information.

As a further complication, recording methods do not cope well with signals that are composed of long strings of 1s or 0s – this is equivalent to trying to record square waves. The way round this is to use a signal of more than 8 bits for a byte, and using a form of conversion table for bytes that contain long sequences of 1s or 0s. A very popular format that can be used is called eight-to-fourteen (ETF), and as the name suggests, this converts each 8-bit piece of code into 14-bit pieces which will not contain any long runs of 1s or 0s and which are also free of sequences that alternate too quickly, like 01010101010101.

All in all, you can see that the advantages that digital coding of audio signals can deliver is not obtained easily, whether we work with tape or with disc. The rate of transmission of data is enormous, as is the bandwidth required, and the error-detecting methods must be very much better and work very much faster than is needed for the familiar personal computers that are used to such a large extent today. That the whole business should have been solved so satisfactorily as to permit mass production is very satisfying, and even more satisfying is the point that there is just one world-wide CD standard, not the furiously competing systems that have made video recording such a problem for the consumer.

For coding TV signals, the same principles apply. Each position on the screen is represented by a binary number which carries the information on brightness and colour. Quite reasonable pictures can be obtained using only eight bits, but for the quality that we are used to a larger number of bits must be used. Once again, the problems relate to the speed at which the information is sampled, and the methods used for digital TV video signals are quite unlike those used for the older system, though the signals have to be converted to analogue form before they are applied to the guns of a CRT. Even this conversion may become unnecessary when CRTs are replaced by colour LCD screens.

A more detailed description and block diagram of CD replay systems is contained in Chapter 13.

Summary

Digital coding has the enormous advantage that various methods can be used to check that a signal has not been changed during storage or transmission. The simplest system uses parity, adding one extra bit to check the number of 1s in a byte. More elaborate systems can allow each bit to be checked, so that circuits at the far end can correct errors. In addition, using coding systems such as eight-to-fourteen can avoid sequences of bits that are difficult to transmit or record with perfect precision.

Chapter 10   Gating and logic circuits

Gates

Digital circuits come in several varieties, but one very important type is the gate, which is used for control actions rather than for counting. A gate in digital electronics means a circuit where the output is a 1 for some specified combination of inputs – this type of circuit is sometimes referred to as a *combinational circuit*. More than 100 years before digital electronics was used, a mathematician called George Boole proved that all of the statements in human logic could be expressed by combinations of three rules which he called OR, AND and NOT.

The importance of this is that if we can provide gate actions corresponding to these three rules, we can construct a circuit that will give a 1 output for any set of logical rules. For example, if we want to have an electric motor switched on when a cover is down, a switch is up and a timer has reached zero or when an override switch is pressed, then this set of rules can be expressed in terms of AND, OR and NOT, and a set of gates can carry out the action.

Figure 10.1 shows the symbols that are used for the three main gate types, the AND, OR and NOT gates, using two-input gates in the example. The action of these gates will be discussed in detail shortly, but for the moment note that the small circle shown in the NOT symbol is used to mean inversion – converting 1 to 0 or 0 to 1. We can combine the other basic symbols with the NOT circle symbol to give symbols for other gate actions. The early digital ICs would typically contain four gates of one type per chip, but modern electronic equipment is more likely to use custom-made chips, with all the gates and their connections formed in one process. Once again, this makes it more useful to show block diagrams rather than gate circuit details. A gate circuit diagram will consist of a large number of gate symbols with joining lines so that the output of a gate will be connected to one or more other gate inputs. Provided that we know what each basic type of gate does, we can analyse the action of complete gate circuits. In this book we are concerned more with block diagrams than with gate circuits, but some knowledge of gate circuits is useful, and in any case, these are closer in spirit to block diagrams than to circuit diagrams.

Logic circuits exist to carry out a set of logic actions such as are used for controls for washing machines, tape recorder drives, computer disc drives and a host of industrial control actions. All logic actions, however complicated, can be analysed into simple actions that are called AND, OR and NOT, so that circuits, called gates, which carry out such actions, are the basis of logical circuits.

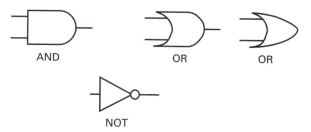

Figure 10.1   Gate symbols for the basic actions of AND, OR and NOT. These are the internationally used symbols and this drawing shows two variants on the OR symbol.

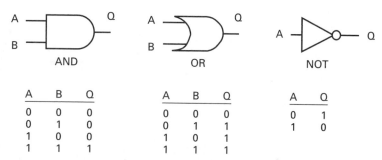

A	B	Q
0	0	0
0	1	0
1	0	0
1	1	1

A	B	Q
0	0	0
0	1	1
1	0	1
1	1	1

A	Q
0	1
1	0

Figure 10.2   Truth tables for gates. The AND and OR gates are illustrated in two-input form.

Remember that ICs are classed as MSI, LSI etc. by the number of equivalent gate circuits.

The action of any gate can be expressed in a *truth table*. This is just a table that shows all the possible inputs to the gate, and the output for each set of inputs. Remember that each input can be 0 or 1 only, so that each input contributes two possible outputs. The total number of outputs is equal to 2^n, where n is the number of inputs. For example, if there are four inputs to a gate, then the number of possibilities is $2^4 = 16$, and a truth table will consist of 16 lines. For a lot of truth tables, there is only one output that is different from the rest, and it is easier to remember which one this is than to try to remember the whole of a truth table.

Figure 10.2 shows truth tables for the basic two-input AND, OR and NOT gates. Of these, the NOT gate is a simple one, with just one input and one output. Its action is that of an inverter. If the input is 0, then the output is not 0, which is 1. If the input is 1, then the output is not 1, which is 0. The other two gate types permit more than one input, and the examples show two inputs, the most common number. The action of the AND gate is to give a 1 output only when both inputs are at 1, and a 0 output for any other combination. The action of the OR gate is to give a 0 output when each input is 0, but a 1 for any other combination of inputs. The same arguments apply to gates with more than two inputs.

NAND and NOR gates

Two particularly useful gate types are made by combining the action of an inverter with that of the AND and OR gates. The combination of NOT and AND gives the NAND gate for which the symbols and truth table (for two

NAND gate

$$A \quad B \quad Q$$

A	B	Q
0	0	1
0	1	1
1	0	1
1	1	0

Figure 10.3 The NAND gate, with symbol, truth table and equivalent circuit.

NOR gate

A	B	Q
0	0	1
0	1	0
1	0	0
1	1	0

Figure 10.4 The NOR gate with symbol, truth table and equivalent circuit.

inputs) are shown in Figure 10.3. The action of this gate is that the output is 0 only when all of its inputs are at 1; which is the action of the AND gate followed by an inverter. The combined action of the OR gate and a following inverter gives the NOR gate, for which the symbols and truth table are shown in Figure 10.4. The output of this gate will be at logic 0 when any one (or more) of its inputs is at logic 1.

There is one further gate that is often used and which is called *exclusive-OR* (XOR). This action, illustrated in Figure 10.5, is closer to what we normally mean by the word 'or', and the output is 1 if either input is 1, but not when both inputs are zero or both are 1. The diagram also shows that the XOR gate is equivalent to the action of a circuit made using an OR, AND and NAND gate combination.

If you would like to read further about gates and other logic circuits, with details of the more advanced methods such as Boolean algebra, take a look at the book *Digital Logic Gates and Flip-flops* from PC Publishing.

A truth table is a simple way of expressing the action of a logic circuit, and the standard gates called OR, AND and NOT can all be illustrated in this way. Gates in IC form often use the NAND and NOR type of gates, equivalent to an AND or OR, respectively, followed by NOT. These inverting gates are easier to produce and the action is often more useful than that of the simpler AND or OR type. The XOR gate is another useful type which gives an action closer to the normal meaning of OR.

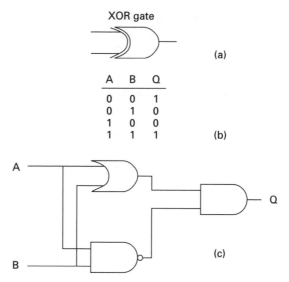

Figure 10.5 The XOR gate with truth table and equivalent circuit.

Analysing gate systems

A circuit that has been made up by connecting several standard gates together, which has several inputs and an output, can be analysed to find what its action is. This analysis can be done by drawing up truth tables, or by a method called Boolean algebra. The truth table method is simpler, but more tedious than the Boolean algebra method, which is not dealt with in this book. The method of analysis by truth table can be summarised in a few rules.

1 Name each input to the circuit (A, B, C) and also each point where the output of one gate is connected to the input of another gate, using different letters for each point. Label the final output as Q.
2 Draw up a blank truth table, using one column for each letter that has been allocated, and with 2^n rows, where n is the number of signal inputs to the circuit.
3 Write in every possible combination of inputs. This is most easily done by starting with 0000 and continuing in the form of a binary count (0001, 0010, 0011, 0100, 0101) up to an input which consists entirely of 1s.
4 Knowing the truth tables for the standard gates, write in the logic states (0 or 1) for the outputs of the gates at the inputs in each line of the truth table.
5 The first set of outputs will now be the inputs for the next set of gates, so that their outputs can be written into the truth table.
6 Continue in this way until the truth table has been completed.

As an example, Figure 10.6 shows a logic diagram for an electronic combination lock. This is a simple design, with four main inputs, and therefore 16 combinations, ignoring the alarm input E. The lock is arranged so that only the correct combination of inputs will open the lock, and any other combination will cause an alarm to sound, so that it cannot be solved by trying each possible combination.

The inputs, A, B, C and D are from switches which are to be set in the pattern needed to open the lock. When these switches have been set, pressing the button E will cause the door to unlock (Q = 1) if the combination

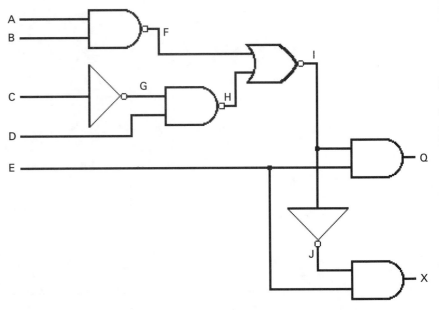

Figure 10.6 A gate circuit used as an example for analysis using truth tables.

is correct, or cause the alarm to sound (X = 1) if the combination is incorrect. To analyse this digital circuit, label the inputs as shown in Figure 10.7.

The important inputs are A, B, C and D, because E is used only after all the others have been set into the correct pattern. The intermediate points, where the output of one gate drives the input of another gate, can now be labelled F, G, H, I, and J as shown in Figure 10.7. Because there are four main inputs, 16 lines of truth table will be needed.

There will be one column for each letter which has been used, but the column for the E input can be placed next to the Q column because the E input is used only when the Q output is decided (after all the other inputs have been set). The logic voltage of E can be written as 1 in each row because the lock will act only when E is set to 1 (the activating button is pushed). All of the possible A, B, C, D inputs can now be written down, starting with 0000, and going through a binary count to 1111, a total of 16 rows in the truth table.

We can now analyse the circuit, Figure 10.8. Inputs A and B are inputs of a NAND gate, where logic is that the output is 0 only when both inputs are 1. The F column, which is the output of this gate, therefore has a zero entered for the last four rows of the table, when both A and B are at logic 1, and a 1 entered for all other rows. The G column is just the inverse of the C column, so that its values can now be written in.

The values in the H column are the outputs of another NAND gate, where the inputs are G and D, so that the output is 0 only when G = 1 and D = 1, as shown. The values in columns F and H are now the inputs to a NOR-gate whose outputs are written into column I.

The logic of the NOR gate is that the output is 1 only when both outputs are at 0, and this occurs only on one line of the table, when A = 1, B = 1, C = 0, D = 1. When I = 1 and E = 1, the output of the AND gate then gives Q = 1, so that the lock opens. For any other combination of inputs at A, B, C and D, the value of Q is 0 and the value in J is 1 (because of the

A	B	C	D	F	G	H	I	J	E	Q	X
0	0	0	0						1		
0	0	0	1						1		
0	0	1	0						1		
0	0	1	1						1		
0	1	0	0						1		
0	1	0	1						1		
0	1	1	0						1		
0	1	1	1						1		
0	0	0	0						1		
0	0	0	1						1		
0	0	1	0						1		
0	0	1	1						1		
0	1	0	0						1		
0	1	0	1						1		
0	1	1	0						1		
0	1	1	1						1		

Figure 10.7 The table with inputs filled in and intermediate values provided for.

A	B	C	D	F	G	H	I	J	E	Q	X
0	0	0	0	1	1	1	0	1	1	0	1
0	0	0	1	1	1	0	0	1	1	0	1
0	0	1	0	1	0	1	0	1	1	0	1
0	0	1	1	1	0	1	0	1	1	0	1
0	1	0	0	1	1	1	0	1	1	0	1
0	1	0	1	1	1	0	0	1	1	0	1
0	1	1	0	1	0	1	0	1	1	0	1
0	1	1	1	1	0	1	0	1	1	0	1
1	0	0	0	1	1	1	0	1	1	0	1
1	0	0	1	1	1	0	0	1	1	0	1
1	0	1	0	1	0	1	0	1	1	0	1
1	0	1	1	1	0	1	0	1	1	0	1
1	1	0	0	0	1	1	0	1	1	0	1
1	1	0	1	0	1	0	1	0	1	1	0
1	1	1	0	0	0	1	0	1	1	0	1
1	1	1	1	0	0	1	0	1	1	0	1

Figure 10.8 The table completed, using the truth tables for the gates to fill in the intermediate values.

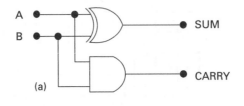

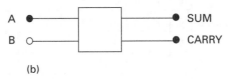

A	B	Sum	Carry
0	0	0	0
0	1	1	0
1	0	1	0
1	1	0	1

Figure 10.9 A half-adder circuit (a) with symbol (b) and truth table (c).

inverter) and the combination of E = 1, J = 1 causes X = 1, sounding the alarm but keeping the door locked. In addition, pressing switch E before any of the others are set will also cause the alarm to sound. The action of this set of gates is to open the lock only for the correct combination of inputs and to sound the alarm for an incorrect combination. Figure 10.8 shows the final state of the truth table.

Any gate circuit can be analysed by drawing up a truth table or a set of truth tables. Though this can be tedious when a circuit has a large number of inputs, it is simpler than the Boolean algebra alternative method. The method relies on using letters to identify all inputs, outputs and intermediate points, and drawing up the truth table in stages, starting with all possible combinations at the inputs.

Arithmetic circuits

The rules of binary addition for two bits are:

Adder circuits

0 + 0 = 0
0 + 1 = 1
1 + 0 = 1
1 + 1 = 0 and carry 1 (the number 10)

The simplest possible adder circuit for binary digits is called a *half-adder*, and it allows two bits to be added, with a main output and a carry bit. The truth table is illustrated in Figure 10.9, and this illustration also indicates that the half-adder can be constructed by using a combination of an XOR gate and an AND gate. The carry bit is zero except when both input bits are 1, which is as required by the rules of binary arithmetic.

This half-adder is such a useful circuit that it is made in IC form in its own right, and it can in turn be used to create other circuits. The name 'half-adder' arises because it can be used only as a first stage in an adder circuit. If we need to add only two bits, the half-adder is sufficient, but if we need to add, say, eight pairs of bits, as when we add two bytes, then the other adders will have three inputs – the bits that are to be added plus the carry bit from the previous stage. For example, adding 1011 and 0011 can use a half-adder for the lowest order pair, giving a zero output and a carry bit. The addition of the next bits is 1 + 1 + 1, because of the carry, and this

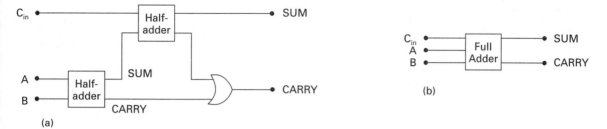

Figure 10.10 The full adder constructed from gates (a) and its symbol (b).

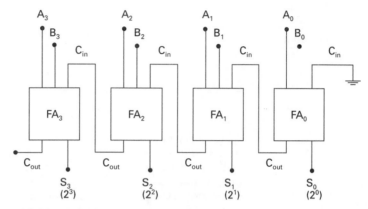

Figure 10.11 A 4-bit parallel adder for two binary numbers.

gives a 1 output and a 1 carry. The next addition uses the carry from the previous stage, and the last addition uses no carry:

Carry	0	0	1	1
A	1	0	1	1
B	0	0	1	1
Sum	1	0	0	0

If we have the half-adder in IC form, we can connect two half-adders along with an OR gate to provide a full adder, which allows a carry input as well as a carry output. Figure 10.10 shows the circuit, its truth table, and the block representation.

A full-adder can be used as a half-adder if the carry input is connected permanently to the 0 voltage level.

In practice, adders can be serial or parallel. A serial adder works on each pair of bits (and any carry) at a time, adding the figures much as we do with pencil and paper. The only complication is the need to store the bits, and we shall see how this is done in Chapter 11. A parallel adder will have as many inputs as there are bits to add, and the outputs will appear as soon as the inputs are present. This is much faster, and though the circuit is more complicated, this is no problem when it can be obtained in IC form. Figure 10.11 shows what is involved in a typical parallel adder circuit for two

Figure 10.12   The half-subtractor in gate form (a) and as a symbol (b).

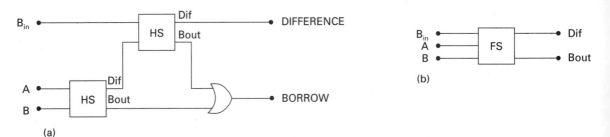

Figure 10.13   The full subtractor in gate form (a) and as a symbol (b).

4-bit binary numbers. In this circuit, all of the inputs are applied at the same time, and the only delay is due to the carry bits, because the addition at each of the later stages is not complete until the carry bit has been generated at the previous stage. This typically takes a few nanoseconds per stage, and the process of passing the carry bit from stage to stage is called *rippling through*.

Since there is never a carry input at the least significant stage, this could use a half-adder, but when the circuit is made in IC form, a full adder with the first carry set to zero is normally used.

Subtraction

The rules for binary subtraction are as follows:

$0 - 0 = 0$
$0 - 1 = -1$ (borrow 1)
$1 - 0 = 1$
$1 - 1 = 0$

and in this table, the −1 entry represents a borrow (or carry back) action from the next higher bit, rather than the carry forward operation that is used for addition.

Figure 10.12 shows how a half-subtractor circuit can be constructed using a NOT gate added to the simple half-adder circuit. Once again, the 'half' reminds you that this circuit does not provide for a borrow, and a 2-bit full subtractor circuit can be made using half-subtractors as shown in Figure 10.13. A single stage like this can be used for serial subtraction.

As you might expect, a full subtractor can be made in parallel form, and an example which provides for two 4-bit numbers, is illustrated in Figure 10.14 using full subtractor units. This, however, can also be built using full adder circuits, using a method of subtraction that we need not worry about at the moment (twos complement with end-around carry). The point is that it allows subtractors to be made using the same IC components as adders, making production more economical.

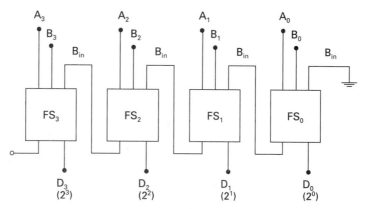

Figure 10.14 A 4-bit full parallel subtractor.

Multiplication and division

The rules of binary multiplication, in table form are:

$0 \times 0 = 0$
$0 \times 1 = 0$
$1 \times 0 = 0$
$1 \times 1 = 1$

and the rules for division are:

$0 \div 1 = 0$
$1 \div 1 = 1$

There are only two rules, because division by zero is meaningless.

Multiplication and division cannot be carried out by gates alone, because they must be done stage by stage. Consider, for example, a multiplication carried out in the same way as we would multiply two denary numbers:

```
      1101
×     1010
      ‾‾‾‾
      0000
     1101.
    0000..
   1101...
   ‾‾‾‾‾‾‾
  10000010
```

This starts by multiplying each bit in the upper number by the lowest-order bit in the second number, producing in this example a set of zeros. Each bit in the second number (the multiplier) will result in a set of four bits (in this example) for the result, and each successive set is shifted left by one place. This action of storage and shifting requires the use of registers, see Chapter 11, and this is followed by an addition, in this case of four 4-bit numbers, to find the final result, which may contain a carry to an extra bit position. Division requires the same types of actions and follows the methods used for long division of denary numbers (once familiar to 8-year-olds, but now a lost art in schools!).

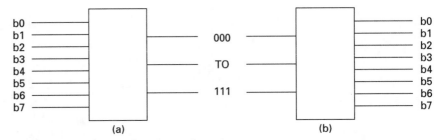

Figure 10.15 Encoder (a) and decoder (b) in block form. The illustration shows 8-to-3 encoding and 3-to-8 decoding.

Arithmetic can be carried out using gates. The simpler addition and subtraction actions can be carried out using only gates, but the actions of multiplication and subtraction require storage and shifting and are dealt with using registers. These actions can be used in serial form (repeating the action for each position in a binary number) or in parallel form (with as many arithmetic units as there are bits in the binary numbers).

Code changing

The need to use codes for digital information often leads to problems in transmitting the digital signals, because we very often cannot provide a signal line for each code signal. Suppose, for example, that we have a set of eight switches, any one of which can be ON (1) or OFF (0), but not allowing more than one switch to be ON at a time. Suppose also that we need to pass this information to some unit which might be a controller or some type of recorder. One obvious method is to use eight lines, one for each switch. A less obvious method is to use the binary numbers for zero to seven (a total of eight codes), which cover the range 000 to 111, using only three lines. A circuit which carries out this change from using eight lines to using three lines is called an 8-to-3 *encoder*.

The word *multiplexer* is often used interchangeably with *encoder*, though strictly speaking the two are not identical. The opposite actions are carried out by the *decoder* or *demultiplexer*.

An encoder is used to convert a set of $2n$ signal lines, on which only one line at a time can be at logic 1, into binary signals on n lines. A decoder performs the conversion from n lines of binary signals to $2n$ lines, of which only one at a time can be at logic 1.

A multiplexer (Mux) is used to select one signal from signals on n input lines and produce the selected signal as an output on one line. A demultiplexer (Demux) has a single input which varies with time, and will provide outputs on n lines.

Figure 10.15 shows an encoder block and truth table for the example of 8-to-3 transformation. A decoder for three lines to eight, also shown here, would work in the opposite way, with a 3-bit binary input and a set of eight lines out, only one of which could be at level 1 for each combination of input bits. You might, for example, use the outputs from the decoder to operate warning lights, if only one at a time needed to be illuminated.

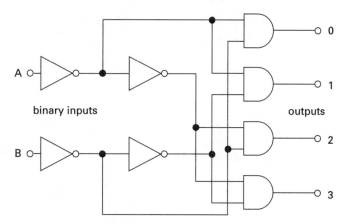

Figure 10.16 A decoder in gate form for 2-to-4 decoding.

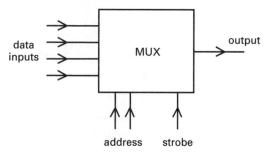

Figure 10.17 A multiplexer in block form, in this example a 4-to-1 MUX with two address lines and a strobe line.

For any decoder, the binary number input is often referred to as an *address number*.

An encoder is needed when each of a set of inputs must generate a unique code pattern. For example, imagine a set of 16 keys (a *hexadecimal keypad*) for which each of the 16 input key characters must generate a unique 4-bit output code – we assume that only one key is pressed at a time. This encoder would require 16 inputs, only one of which is selected at any one time, and would provide four output lines to provide the parallel bit pattern. If we needed to reconstitute the pattern of lines at the input, we would use a decoder which would have four input lines and would use the bits on these lines to select one and only one of 16 output characters. Encoders and decoders can be made using gates, and Figure 10.16 shows an example of a 2-to-4 line decoder using inverters and AND gates, along with a truth table analysis of the action.

Now let us look at a multiplexer. Figure 10.17 shows in block form what a 4-to-1 multiplexer is expected to do. There are four data inputs, two address inputs and a strobe input. The address inputs are used to select one of the four data lines, using an encoding action (which is why encoders and multiplexers are often taken to be almost identical). The data on the selected line is transferred to the single output line when the signal on the strobe line is low (logic 0). This allows all of the data to be sent down one line, a bit at a time, by establishing an address number, pulsing the strobe input, and

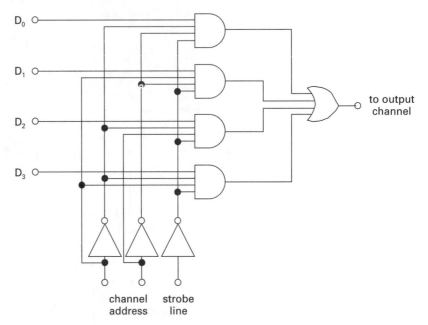

Figure 10.18 A 4-to-1 multiplexer in gate form.

repeating this for another address number until you have cycled through all the address numbers.

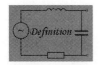

A *strobe* pulse is one that is enabled at a set time. The name was originally used in radar to mean a pulse that was used to enable the receiver for a time that excluded nearby targets.

This type of action, allowing a mass of data to be selected for transmission down a single line, is called *time-division multiplexing*. The use of a carrier modulated in more than one way (such as the colour TV sub-carrier) is another form of multiplexing.

Figure 10.18 shows a gate circuit that can carry out the action of multiplexing a set of four inputs. For four inputs, a 2-bit address number is needed to select which input will be connected to the output line when the strobe input goes to logic 0. The opposite action of a demultiplexer is shown in block form in Figure 10.19. There is a single data input and a strobe input, along with the address inputs, in this case using two lines for the addresses of four output lines. Once again, the data on the input line will appear at the selected output line when the strobe pulse voltage is low, and the truth table is as follows.

The selected output data line in this example takes the logic value of zero (the same as the data line input) when the strobe appears.

Encoders, decoders, multiplexers and demultiplexers are used to change the way that digital codes are carried. Usually, one of the codes will be 8–4–2–1 binary. The distinguishing feature of the multiplexer is that it allows any

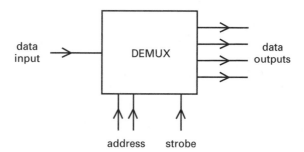

Figure 10.19 Block for a 1-to-4 demultiplexer.

Table 10.1

Address		Data	Strobe	Outputs			
A	B			Y_0	Y_1	Y_2	Y_3
0	0	0	0	0	1	1	1
0	1	0	0	1	0	1	1
1	0	0	0	1	1	0	1
1	1	0	0	1	1	1	0

number of input lines to be used to place signals on a single output line, controlled by a selection address number and an enabling strobe pulse. The demultiplexer carries out the opposite action, assuming that the strobe pulses are correctly synchronised.

Chapter 11   Counting and correcting

Sequential circuits

Gate circuits are one basic type of digital circuit, the combinational circuit. The other basic type of digital circuit is the *sequential* type. The output of a sequential circuit, which may mean several different output signals on separate terminals, will depend on the *sequence* of inputs to the circuit. One simple example is a counter IC, in which the state of the outputs will depend on how many pulses have arrived at the input.

The basis of all sequential circuits is a circuit called a *flip-flop*, and the simplest flip-flop circuit is called the S–R flip-flop, with the letters S and R meaning Set and Reset. Though this type of flip-flop is not manufactured in IC form (because it is just as simple to manufacture more complex and more useful types), it illustrates the principle of sequential circuits more clearly than the more complicated types.

A flip-flop is a circuit where the output(s) change state for some set of inputs and which will remain unchanged until another set of inputs is used. Unlike a gate, simply changing inputs does not necessarily change outputs.

One feature of flip-flops is the use of two outputs, conventionally labelled as Q and Q# (or $\bar{Q}$). The Q output is the main output, and the Q# output is the inverse of the Q output.

Figure 11.1 shows the block for one variety of simple S–R flip-flop along with a table of inputs and outputs. The notable point is that the table shows five lines, and that the line which uses 0 for both inputs appears twice. The output for S = 0 and R = 0 can be either Q = 1 or Q = 0, and the value of Q depends on the sequence that leads to S = 0, R = 0. If you use inputs S = 1, R = 0, then Q = 1, and this value will remain at 1 when the input changes to S = 0 and R = 0. Similarly, if the inputs are S = 0, R = 1, the Q output becomes 0, and this remains at zero when the inputs change to S = 0, R = 0. One state, with S = 1 and R = 1, is forbidden because all the uses for flip-flops require Q# always to be the opposite of Q, and the S–R flip-flop violates this requirement when S = 1 and R = 1.

The meanings of S and R are now easier to understand. Using S = 1 and R = 0 will *set* the output, meaning that Q changes to 1. Using S = 0, R = 1 will *reset* the output, meaning that Q changes to 0. The input S = 0, R = 0 is a storing input, making the flip-flop remain with the output that it had previously. Though the single S–R flip-flop is not manufactured in IC form, it is useful (as a latch, see later) and can be constructed using two NAND gates as shown in Figure 11.2. Multiple S–R latch chips are available.

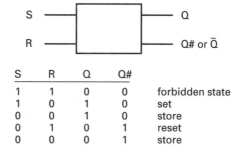

S	R	Q	Q#	
1	1	0	0	forbidden state
1	0	1	0	set
0	0	1	0	store
0	1	0	1	reset
0	0	0	1	store

Figure 11.1 The S–R flip-flop with its state table.

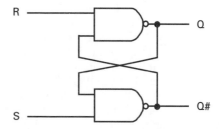

Figure 11.2 How a S–R flip-flop can be constructed using NAND gates. Another version can be made using NOR gates, with a slightly different state table.

Sequential circuits are the other important digital type, using in counting and for memory actions. The simplest type is the S–R flip-flop (or latch), whose output(s) can be set by one pair of inputs and reset by reversing each input. The output is held unchanged when both input signals are zero.

The J–K flip-flop

Though simpler types exist, the most important type of flip-flop is that known as the Master–Slave J–K flip-flop, abbreviated mercifully to J–K. This is a clocked circuit, meaning that the action of the IC is carried out only when a pulse is applied to an input labelled *clock*. This is the type of flip-flop that is most commonly manufactured in IC form.

Using this type of flip-flop allows the actions of a number of such circuits to be perfectly synchronised, and avoids the kind of problems that can arise in some types of gate circuits when pulses arrive at slightly different times. These problems are called *race hazards*, and their effect can be to cause erratic behaviour when a circuit is operated at high speeds. When clocking is used, the circuits can be operated synchronously, meaning that each change takes place at the time of the clock pulse, and there should be no race hazards.

The J–K flip-flop uses three main inputs, labelled as J, K and clock (Ck). Of these, the J and the K are programming inputs where the voltage levels will control the action of the flip-flop at the time when the clock pulse triggers the circuit. Because there are two programming inputs, and each of them can be at either of two levels (0 or 1), there are four possible modes of operation of the J–K flip-flop.

We can describe each of those modes of action in terms of what voltage changes occur at the output when the clock-pulse edge arrives. The usual triggering edge is the trailing back edge, the level 1 to level 0 transition of

Table 11.1 A J–K flip-flop and outputs

J	K	Q_n	Q_{n+1}	comment
0	0	0	0	no change, latched
0	0	1	1	no change, latched
0	1	0	0	reset on clock
0	1	1	0	reset on clock
1	0	0	1	set on clock
1	0	1	1	set on clock
1	1	0	1	toggle
1	1	1	0	toggle

a pulse, and on Table 11.1, the condition of the Q output before the arrival of trailing edge is indicated as Q_n and the condition after the edge as Q_{n+1}.

The input is the clock pulse, with the J and K terminals used for programming, and outputs are taken from the Q and Q# terminals. Remember that the Q# output is *always* the inverse of the Q output – in older texts this is often shown by drawing a bar over the Q. The table shows the possible states of inputs before and after the clock pulse, showing how the voltages on the J and K pins will determine what the outputs will be after each clock pulse.

The four modes which are indicated in Table 11.1 are as follows:

1 The *hold* mode. With J = 0 and K = 0, the output is unchanged by clock pulses. This avoids the need for further gating if you want to keep the flip-flop isolated.
2 The *set* mode, with J = 1, K = 0. Whatever the state of the outputs are before the clock pulse edge, Q will be set (to 1) at the triggering edge and will stay that way.
3 The *reset* mode, with J = 0, K = 1. Whatever the state of the outputs before the triggering edge, Q will be reset (to 0) at the edge and will stay that way.
4 The *toggle* mode, with J = 1, K = 1. The output will change over each time the circuit is clocked.

In addition to the inputs J and K which affect what happens when a clock pulse comes along, these flip-flops are usually equipped with S and R terminals (which are also called preset and reset terminals). These will affect the Q and Q# outputs of the flip-flop immediately, irrespective of the clock pulses. For example, applying a zero input to the R input will reset the flip-flop whether there is a clock pulse acting or not. This action is particularly useful for counters which have to be reset so as to start a new count.

The J–K flip-flop uses two programming inputs labelled as J and K, where the voltage determines how the flip-flop will operate. One particularly useful input is J = 1, K = 1 which allows the flip-flop to reverse (or *toggle*) the output state at each clock pulse, so that its output is at half the pulse rate of the clock. Two inputs labelled S and R (or Pr and R) allow the output to be set (preset) or reset independently of the clock pulses.

The use of two programming inputs, J and K, therefore causes a much more extensive range of actions to be available. Note that if both the J and K

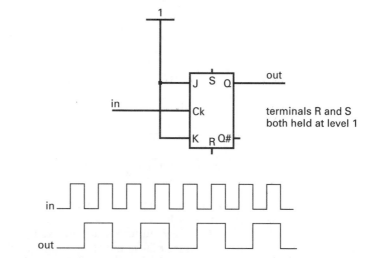

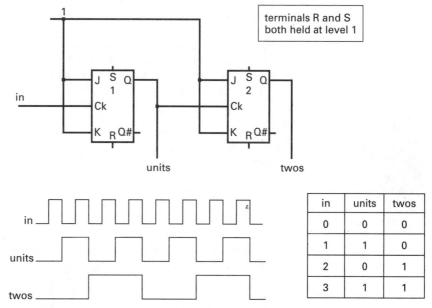

Figure 11.3 Using a J–K flip-flop as a counter or frequency divider.

in	units	twos
0	0	0
1	1	0
2	0	1
3	1	1

Figure 11.4 A two-stage simple counter using J–K flip-flops.

terminals are kept at level 1, the flip-flop will toggle so that the output changes over at each clock pulse, as is required for a simple binary counter, see Figure 11.3.

Flip-flops are the basis of all counter circuits, because the toggling flip-flop is a single stage scale-of-two counter, giving a complete pulse at an output for each two pulses in at the clock terminal. By connecting another identical toggling flip-flop so that the output of the first flip-flop is used as the clock pulse of the second, a two-stage counter is created, so that the voltages at the Q outputs follow the binary count from 0 to 3, as Figure 11.4 shows. This principle can be extended to as many stages as is needed,

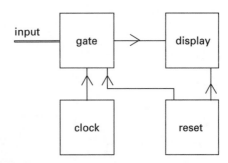

Figure 11.5 Simplified block diagram for a frequency meter.

and extended counters of this type can be used as timers, counting down a clock pulse which can initially be at a high frequency.

The type of counter which uses toggling flip-flops in this way is called *asynchronous*, because the clock inputs to the flip-flops do not occur at the same time and the last flip-flop in a chain like this cannot be clocked until each other flip-flop in the chain has changed. For some purposes, this is acceptable, but for many other purposes it is essential to avoid these delays by using *synchronous* counter circuits in which the same input clock pulse is applied to all of the flip-flops in the chain, and correct counting is assured by connecting the J and K terminals through gates. Circuits of this type are beyond the scope of this book, and details can be found in any good text of digital circuit techniques. In this book we shall, as usual, concentrate on block diagrams which are not concerned with whether a counter is synchronous or not.

The toggling flip-flop is the basis of all counter circuits, which use the clock as an input and the Q outputs as the outputs. In such a circuit, the Q outputs provide a binary number which represents the number of complete clock pulses since the counter was reset.

Counter uses

Apart from their obvious applications to calculators, counters are used extensively in electronics, and particularly in modern measuring instruments. ICs exist to provide for measurement of all the common quantities such as DC voltage, signal peak and RMS amplitude, frequency, etc., using counting methods. Most of the modern ICs of this type provide for digital representation, so that the outputs are in a form suitable for feeding to digital displays. The simplest application of counters to instruments is for frequency measurement.

Frequency meter

A frequency meter makes use of a high-stability oscillator to provide clock pulses, normally crystal controlled. In its simplest form, the block diagram is as shown in Figure 11.5, with the unknown frequency used to open a gate for the clock pulses. The number of clock pulses passing through the gate in the time of one cycle of the input will provide a measure of the input frequency as a fraction of the clock rate. For example, if the clock rate is 10 MHz and 25 pulses of the unknown frequency pass the gate, then the input frequency is 10/25 MHz, which is 400 kHz. A counter circuit can find the number and display it in terms of the clock frequency to give a direct reading of the input frequency.

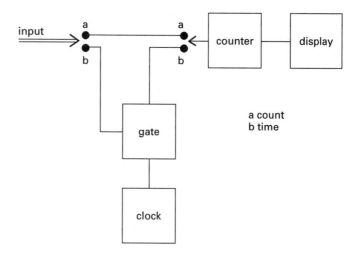

Figure 11.6 Block diagram for a counter/timer instrument.

In this simple form, the frequency meter cannot cope with input frequencies that are greater than the clock rate, nor with input frequencies that would be irregular submultiples. For example, it cannot cope with 3.7 gated pulses in the time of a clock cycle. Both of these problems can be solved by more advanced designs. The problem of high frequencies can be tackled by using a switch for the master frequencies that allows harmonics for the crystal oscillator to be used. The problem of difficult multiples can be solved by counting both the master clock pulses and the input pulses, and operating the gate only when an input pulse and a clock pulse coincide. The frequency can then be found using the ratio of the number of clock pulses to the number of input pulses.

Suppose, for example, that the gate opened for 7 input pulses and in this time passed 24 clock pulses at 10 MHz. The unknown frequency is then $10 \times 7/24$ MHz which is 2.9166 MHz. Frequency meters can be as precise as their master clock, so that the crystal control of the master oscillator determines the precision of measurements.

Counter/timer

A counter/timer is used for the dual roles of counting pulses and measuring the time between pulses, and Figure 11.6 shows the block diagram, omitting reset and synchronising arrangements.

The counter portion is a binary counter of as many stages as will be required for the maximum count value. The input to the counter is switched, and in the counting position, the input pulses operate the counter directly. When the switch is changed over to the timing position, the input pulses are used to gate clock pulses, and it is the clock pulses that are gated. In a practical circuit, the display would be changed over by the same switch so as to read time rather than count.

Frequency meters and counter/timer circuits are obvious applications of counter circuits to measuring devices, and they have completely replaced older methods.

Summary

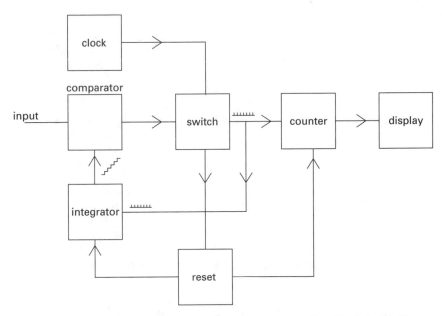

Figure 11.7 Block diagram for a digital voltmeter. The block is for the main measuring unit, neglecting resistors used to extend the range for larger voltages.

Digital voltmeter

At one time, all voltmeters were of the analogue type. The voltage that was to be measured was used to pass current through a resistor of precise and known values, and the current passing was measured by a meter. The meter scale was calibrated in terms of volts, so that the reading was direct; it did not require any calculations. The problem of this type of meter was that it took a current from the circuit, so that it altered, even if only slightly, the voltage that it was measuring. Modern digital meters take a negligible current and present the results as a number rather than as a scale reading.

The most common IC of this type is the digital DC voltmeter IC, and a brief account of its action provides some idea of the operating principles of many instrumentation ICs.

Referring to the block diagram of Figure 11.7, the voltmeter contains a precision oscillator that provides a master pulse frequency. The pulses from this oscillator are controlled by a gate circuit, and can be connected to a counter. At the same time, the pulses are passed to an integrator circuit that will provide a steadily rising voltage from the pulses.

When this rising voltage matches the input voltage exactly, the gate circuit is closed, and the count number on the display then represents the voltage level. For example, if the clock frequency were 1 kHz, then 1000 pulses could be used to represent 1 V and the resolution of the meter would be 1 part in 1000, though it would take one second to read one volt. After a short interval, usually around 0.25 second (determined by using another clock or a divider from the main clock) the integrator and the counter are reset and the measurement is repeated. Complete meter modules can be bought in IC form, using fast clock rates and with the repetition action built in.

Summary

The digital voltmeter is a less obvious use of counters in measurement, and its action depends on the use of a very precise integrator in IC form to produce a sawtooth voltage waveform from a set of counted pulses.

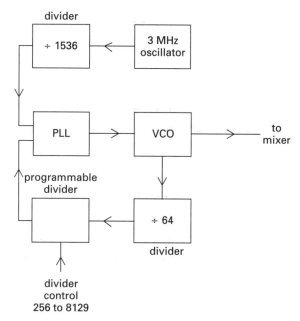

Figure 11.8 A frequency synthesiser form of oscillator as used for TV tuners.

A very different application of digital methods is illustrated in Figure 11.8 which shows a *frequency synthesiser* such as is used on TV receivers and in a simpler version for FM receivers as a local oscillator for the superhet action. This, strangely enough, depends for its action on an *analogue* circuit called a phase-locked loop (PLL) which compares the phases of two input signals of about the same frequency and where the output is a voltage the size of which depends on the amount of phase difference between the input signals. If the input frequencies are different, the output of the PLL is not a steady voltage, but if the frequencies are identical but with a phase difference, then the PLL output is a steady voltage.

The reference oscillator uses a crystal control and is precise enough to maintain a frequency as constant as any of the transmitted carriers. Its frequency is typically 3 MHz, which allows the use of comparatively low-cost components. This 3 MHz frequency is divided down by a factor of 1536 to give 1953.125 Hz, and this frequency is used as one input to the PLL. Another oscillator, labelled as VCO (voltage controlled oscillator) is used as the local oscillator for the TV receiver, and will be working at a UHF frequency, such as 670.75 MHz. The point of using the VCO is that its operating frequency can be altered by applying a steady input voltage, replacing any form of mechanical tuning. This oscillator frequency is divided by a fixed factor of 64 to give 10.480468 MHz, and this frequency is then used as the input of a programmable divider, where the division ratio can be anything from 256 to 8191. The numbers that control these division ratios are obtained from a memory within the tuner, and this is programmed with a control number for each possible channel when the receiver is first set up.

For example, if the frequency of 10.480468 MHz is divided by 5366, the result is 1953.125 Hz, which is identical to the frequency from the crystal-controlled oscillator. If the VCO frequency changes slightly, the frequencies into the PLL will no longer be equal, and a voltage will be generated from the PLL and used to correct the VCO. If you want to change channels, you

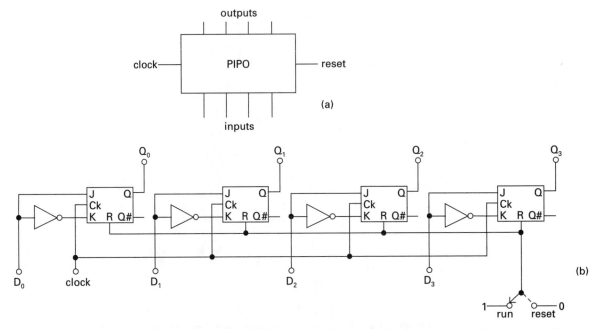

Figure 11.9 The PIPO register, showing the symbol (a) and a circuit using flip-flops (b).

press a switch that alters the divider ratio, causing a change in the PLL output that will then alter the frequency of the VCO until the new channel is perfectly tuned. It's all a long way from the tuning of the early radios.

Frequency synthesisers are another application of counter circuits, and this type of action is now extensively used in radio and TV receivers to produce the correct oscillator frequency for a superhet receiver circuit. This type of tuning is better able to use remote control and push-button selection actions, and has completely replaced the old style rotating-knob tuning for TV receivers.

Registers

The use of flip-flops in binary counters is just one application of these versatile building-blocks. A register, usually found in integrated form, is another method of using flip-flops for storage of binary digits and for logic operations that are called shifting and rotating.

A register is created from a set of flip-flops connected together, and there are four basic methods which are distinguished by the initials of their names. The simplest type is called parallel-in parallel-out (PIPO), and a register of this type has no connections between the flip-flops, Figure 11.9. Each flip-flop has an input and an output, and the principle is that the inputs will determine the outputs, and these outputs will be maintained until another set of input signals are used to change them. The use of a register of this type is for storage, because each flip-flop is storing a bit (0 or 1) and will hold this bit until new input signals are used or until the power is switched off. This system was at one time used for computer memory, but it requires too much power (because each flip-flop draws current whether it is storing 1 or 0) for modern systems. The system, also called static RAM (SRAM) is still used for small fast memories.

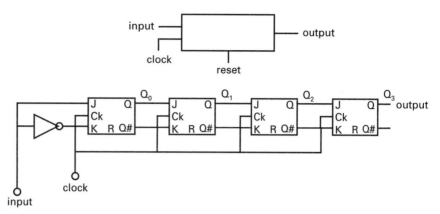

Figure 11.10 The SISO register with symbol and a circuit using flip-flops.

The use of a register for very short-term storage is called *latching*. For example, using a latch register allows a set of digits which may exist for only one clock cycle (the time between two clock pulses), to be viewed on a display for as long as is needed. This facility is particularly useful for measuring instruments in which the signals may change rapidly but the display must remain static for long enough to be read. A timer, for example, would be of little use if the time, measured in units of milliseconds, was displayed directly as the counter output, because the display on the least significant digits would be constantly changing and would not be readable until the timer stopped.

It is better to use a register as a latch between the counter and the display so that the time can be displayed without stopping the counter – this allows intermediate times to be displayed. We can, for example, display the time for one lap of a race, while allowing the counter to continue timing the rest of the race. A latching action is also an essential part of a modern digital voltmeter (designs more modern than the illustration of Figure 11.7), providing a steady reading (which is updated once per second) even when the voltage level is fluctuating slightly.

A register is a circuit that uses a set of flip-flops. The simplest type is the parallel-in parallel-out register (PIPO) or latch, in which the flip-flops are independent (apart from a reset line), each with an input and an output. The action is used mainly for storing the bits of a digital signal, so that the register must use as many flip-flops as there are bits in the stored signal.

The opposite type of register is titled serial-in serial-out (SISO). Each flip-flop has its output connected to the input of the next flip-flop in line, Figure 11.10, so that the whole register has just one input and one output. This is used as a signal delay and as a counter. On each clock pulse input, the bits in the flip-flops are shifted, meaning that they cause the next flip-flop in line to change, and for a set of four flip-flops this will cause the waveforms to appear as shown in Figure 11.11. If there are no connections to the intermediate flip-flop outputs, this register acts as a divide-by-four or count to four circuit.

The parallel-in serial-out (PISO) register is used mainly to convert parallel signals on a set of lines into a serial signal on one line. A set of bits on

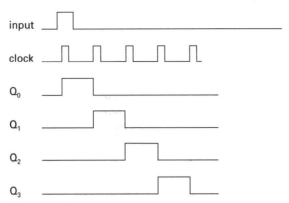

Figure 11.11 The result of using a square pulse input for the first of a set of clock pulses. The pulse is cycled through the register by one stage on each clock pulse.

the input lines will affect each flip-flop in the register, and by pulsing the clock line once for each flip-flop, each bit is fed out in turn from the serial output. The opposite type is the serial-in parallel-out (SIPO) register in which bits are fed in one at a time on each clock pulse and appear on the parallel output after a number of clock pulses that is equal to the number of flip-flops in the register.

You never need to make registers from individual flip-flops, because all forms of registers are obtainable in completely integrated form, often as units that can be connected so as to obtain whatever action you want. Integrated registers often feature shift direction controls so that by altering the voltage on a pin, you can switch between left-shift and right-shift. When the input of a serial register is connected to the final output, the action is *rotation* – the bits are cycled round one stage for each clock pulse.

The registers with serial input or output are used extensively for converting between serial and parallel data signals, and are an alternative to the use of multiplexers and demultiplexers (see Chapter 10).

Error control and correction

Any system that uses digital signals is much less likely to be upset by noise than the equivalent analogue signal. Consider, for example, the square-wave signal in Figure 11.12 which has been degraded by noise, and which can be 'cleaned up' simply by slicing the received signal at the two extremes of voltage levels. This type of cleaning up is impossible when an analogue signal is being used, because there is no simple way of distinguishing the noise from the analogue signal itself.

Figure 11.13 is a graph which shows how these two types of signals behave when the signal-to-noise ratio decreases by the same amounts. The analogue system is affected progressively, but the digital system is almost unaffected by the noise until a point is reached when the system suddenly crashes as the *bit error rate* (BER) rises. This does not mean that digital methods cannot be used when the noise level is high, because there are a number of ways in which the information can be accurately recovered even from signals with a high BER. These methods are always used in digital recording systems (such as CD) and for transmission systems (digital telephone links) and are the main reason for the widespread adoption of digital

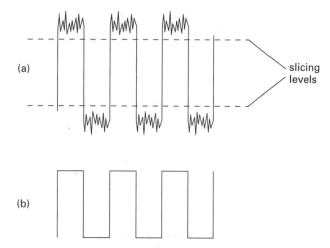

Figure 11.12 Noise immunity of a digital signal. The noise affects the tips of the waveform, which can be squared by clipping (slicing) at suitable levels.

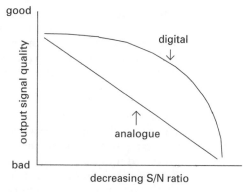

Figure 11.13 Comparison of noise immunity for comparable analogue and digital circuits.

methods for signalling. A typical bit error rate for noisy conditions is 1 error in 1000 (a BER of 10^{-3}), and for better conditions, about 1 in 10 000 (BER = 10^{-4}).

Summary

All forms of electronic signals are degenerated by noise. For analogue signals, this is expressed as the signal/noise (S/N) ratio, and this figure should be high, typically 40 dB or more. For digital signals, the effect is expressed as bit error rate (BER), and this should be low, of the order of 1 in 1000 (10^{-3}) or 1 in 10 000 (10^{-4}) if data correction methods are to work well. The important difference between analogue and digital methods is that whereas analogue signals gradually deteriorate, digital signals are almost unaffected until the noise reaches a critical level, when the digital signal becomes totally garbled.

The simplest type of system that can be used for correcting signals sent from a transmitter to a receiver is known as Automatic ReQuest for repeat

(ARQ). A standard code (ASCII) has for many years been used for tele-printers and computers – the letters of the name stand for American Standard Code for Information Interchange. This uses two special code numbers that are referred to as ACK and NAK. If a distant receiver detects that a code pattern has no errors, it transmits back the ACK (Acknowledge) code. If errors have been detected, the NAK (Negative acknowledge) code is returned, and this is used to make the transmitter repeat the transmission of the last block of signal code.

A more advanced system is known as forward error control (FEC), and we have looked already at one version of this in Chapter 9, the use of parity bits. All FEC systems make use of the addition to the signal bits of extra redundant bits which, when suitably processed, are capable of identifying and in some cases, correcting, the errors.

The main causes of bit errors in digital signals are *white noise* and *impulsive noise*. White noise consists of a mixture of frequencies that is fairly evenly spread over the frequency range; no frequency is affected more than any other. This type of noise produces errors that are at random. Impulsive noise, such as is produced by car ignition systems, is at some definite pulse rate and it can create bursts of errors on each pulse. The errors that can occur are of three types: detectable and correctable, detectable but not correctable, and undetectable (and therefore uncorrectable).

If an error can be detected but not corrected, it is possible that the error can be concealed in some way so that it causes less damage. For example, the system could ignore the error in a number and treat it as a zero level, or it could make use of the last known correct value again or it could calculate an average value based on the previous and next correct values. This last method is called *interpolation*.

Using a check sum

The check sum is an error control scheme which is often used for digital recording systems, particularly magnetic disc and tape. The data must be organised into long 'blocks', each of which can be addressed by a reference. For each block of digital data (numbers) the digital sum of the numbers, the *check sum*, in a block is stored at the end of that block. When the data is recovered, the check sum can be recalculated from the block of replayed numbers and compared with the original to find if any errors have occurred during reading or storage.

The simple check-sum technique can only identify when an error has occurred and can not indicate where in the block the error is located. A modification of this method is called the weighted check-sum method, and it relies on the behaviour of prime numbers (number which have no factors other than 1), such as 1, 3, 5, 7, 11, 13 etc. In a weighted check-sum system, each data number in a block is multiplied by a different prime number, and the check sum is calculated from this set of numbers. If a different check sum is found on replay, the difference between the calculated and the stored check sum is equal to a prime number, the same prime number as was used to multiply the data number that has been corrupted, so locating the number. The mathematical basis of this system is beyond the scope of this book.

The use of ACK and NAK codes along with parity is a simple system for correcting transmitted data, but nowadays the use of systems such as check sums is more usual. The more advanced systems deal with a block of data numbers at a time, and can detect an error in the block, and correct this error if the coding allows its position to be located.

Hamming and other codes

Hamming codes are a form of error correcting codes that were invented by R. W. Hamming, the pioneer of error control methods. The details of Hamming codes are too mathematical for this book, but the principles are to add check bits to each binary number so that the number is expanded – for example, a 4-bit number might have three check bits added to make the total number size seven bits. The check bits are interleaved with the number bits, for example using positions 1, 2 and 4 for check bits, and positions 3, 5, 6 and 7 for the bits of the data number.

When an error has occurred, the parity checking of the check bits will produce a number (called a *syndrome*) the value of which reveals the position of an error in the data. This error can then be corrected by inverting the bit at that position. A simple Hamming code will detect a single bit error, but by adding another overall parity bit, double errors can be detected. Hamming codes are particularly useful for dealing with random errors (caused by white noise).

The cyclic redundancy check (CRC) method is particularly effective for dealing with burst errors caused by impulsive noise, and is extensively used for magnetic recording of data. Once again, the mathematical details are beyond the scope of this book, but in general, a data block is made up from the data word, a generator code word and a parity check code word. The data word is divided by the generator word to give the parity check word and the whole block is recorded. On replay, the generator code is separated, and the whole block number is divided by this code. If the result is zero, there has been no error. If there is a remainder after this division, the error can be located by using this remainder along with the parity code.

The basic Hamming and CRC systems have been improved and developed over the years to produce systems that are more effective in dealing with both random errors and bursts of errors. These systems include BCH codes for random error control, Golay codes for random and burst error control and Reed–Solomon codes for random and very long burst errors with an economy of parity bits. The Reed–Solomon coding system is used for CD recording and replay.

The more advanced methods of error detection and correction, ranging from simple Hamming codes through CRC to Reed–Solomon all make use of added bits in a block of data, and mathematical methods that allow the position of an error to be found. These methods also allow for error correction.

Chapter 12   Microprocessors, calculators and computers

The micro-processor

A microprocessor is a logic chip that contains gates and flip-flops which are arranged so that the connections between the internal units are controllable. The control is arranged by using a set of bits that form a *program* input to the microprocessor and which are stored in a register (the program register). The microprocessor can *address* memory, which means that it can select an item of stored data from any location in external memory chips and make use of it and it can store results also in a selected portion of memory. Within the microprocessor chip itself, logic actions such as the standard NOT, AND, OR and XOR gate actions can be carried out on a set of bits, as well as a range of other register actions such as shift and rotate, and some simple arithmetic. The fact that any sequence of such actions can be carried out under the control of a program which is also read in from memory is the important item that provides the definition of a microprocessor.

Figure 12.1 shows a block diagram which is extensively used for teaching purposes, and which is still applicable to many types of microprocessor that are used for industrial control actions. Before we can look at the action we need to know what the units are and how they are connected.

One important point is the use of *bus* connections. A bus is a set of lines, each line used for a single-bit signal, that connects several units. Most buses are bidirectional, meaning that signals can be sent in either direction, and this is achieved by using enabling pulses to the units that are exchanging bits. For example, if a unit is enabled and its outputs are connected to the lines of a bus, then a set of bits will be placed on the bus lines and these bits can be used as inputs for any other unit whose inputs are connected to the bus and which is also enabled. In practice, both the inputs and the outputs of each unit will be connected to the same bus, and the enabling pulses allow either reading (input), writing (output) or disabled (not connected) states. The use of buses allows the construction of the microprocessor to be much simpler than it would be if separate connections had to be used for each set of inputs and outputs.

In use, the main data bus can be connected with other units, such as memory and latches, to allow the microprocessor to make use of these other circuits. The two important buses which are used in this way are the data bus and the address bus, and these, along with control signals, allow the microprocessor to control other units as if they were a part of the micro-processor itself.

A microprocessor is a type of universal logic IC chip which can carry out a set of actions in sequence. The sequence and the actions are controlled by a program, which consists of binary numbers that are used to control gates within the microprocessor. Other units (such as memory) are connected by way of buses, sets of lines which connect to all units. When a bus is used to pass signals, the sending and the receiving units are enabled and other

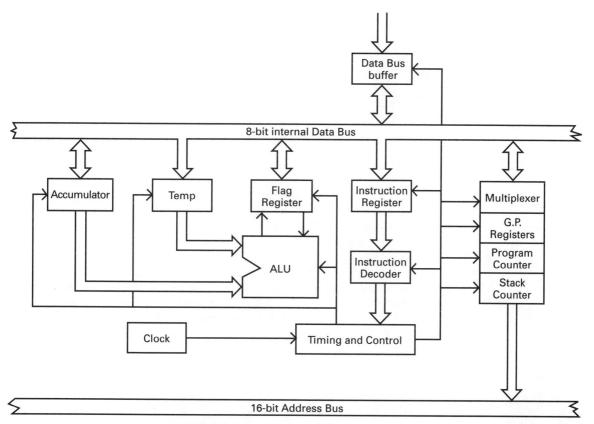

Figure 12.1 A block diagram of the circuits inside an older type of 8-bit microprocessor. Modern chips are much more elaborate, but the basic structure is similar.

units are disabled, allowing the same buses to be used for different signals in either direction.

The action, simplified, is as follows. When the microprocessor starts work, it will read a byte from some fixed address in the external memory store. This first byte is always an instruction, and the microprocessor reads it into an internal register, the *instruction register*, where it is decoded. As a result of reading the instruction code (by matching it with one of a set permanently stored in the microprocessor), an action will be carried out. This action is controlled by a short built-in program, called a microprogram. There is a microprogram for each possible microprocessor action.

This action might require data – for example, if you want to AND two bytes together then these bytes constitute the data, so that a program command for ANDing two bytes would have to be followed by reading the bytes that were to be ANDed. The instruction code must contain the information that allows the microprocessor to find and read the data (in a register or in the memory), and when the data is read the action will be carried out and the result is stored in a register called the *accumulator* (or it can be stored in memory if another instruction provides for this). The next byte that the microprocessor reads will then be another instruction. The only difference between a byte of instruction and a byte of data is the way that they are arranged in the memory, instruction first and its data following, and this correct sequence ensures that the bytes are routed to the correct registers in the microprocessor. The timing of actions is determined by clock pulses which are supplied by a crystal-controlled oscillator, usually at a high speed (typically 25 MHz or more). Each action usually requires more than one clock pulse because of the number of steps in an action, even a simple action like adding.

All actions such as AND or ADD are carried out in the arithmetic and logic unit (ALU), and this is arranged so that one of its inputs is from the accumulator register and its output also is to the same register – remember that the microprocessor is clock controlled, so that input and output occur at different times. This arrangement means that the accumulator always provides one of the bytes in any arithmetic or logic action, and it always holds the result following such an action. If a second input is required, as it is for most actions, it can be supplied from the temporary register which in turn is filled from any unit connected to the internal data bus.

The 'data size' of a microprocessor is stated in terms of the number of bits in the accumulator register. Early microprocessors (still in use) had an 8-bit register and are therefore 8-bit microprocessors. More recent types are 32-bit microprocessors, and the development of 64-bit microprocessors is under way.

All actions start by reading a program code from memory. This code is placed in the program register of the microprocessor and used to activate a microprogram that is stored within the microprocessor itself. This may require further number data to be read from memory, and the programmer must ensure that these numbers follow the program code(s). All microprocessors make use of an accumulator register which can supply one number for an action, and where the result of an action will also be stored. The size of the accumulator in terms of bits is used as a measure of the capability of the microprocessor.

Memory

Memory is a name for a type of digital circuit component which can be of several varieties. The basis of a single unit of memory is that it should retain a 0 or a 1 signal, and that this signal can be connected to external lines when needed for reading (copying) or writing (storing). The method that is used to enable connection is called *addressing*. The principle is that each unit of memory should be activated with a unique combination of signals that is present on a set of lines called the address bus. Since each combination of bits constitutes a binary number, the combinations are called address numbers. We have already looked at this idea in connection with encoders and multiplexers.

There are two fundamental types of memory, both of which are needed in virtually any microprocessor circuit. One type of memory contains fixed unalterable bits, and is therefore called read-only memory, or *ROM*. The important feature of ROM is that it is *non-volatile*, meaning that the stored bits are unaffected by switching off power to the memory and they are available for use whenever power is restored. For example, the microprograms are stored in ROM that is contained within the microprocessor itself.

Since there must be an input to the microprocessor whenever it is switched on, ROM is essential to any microprocessor application, and in some applications it might be the only type of memory that is needed. The simplest type of such a ROM consists of permanent connections to logic 0 or logic 1 voltage lines within a chip, gated to output pins when the correct address number is applied to the address pins.

Read-write memory is the other form of memory which for historical reasons is always known as *RAM* (Random-Access Memory). This is because, in the early days, the easiest type of read-write memory to manufacture consisted of a set of serial registers, from which bits could be read at each clock pulse in sequence. This meant that if you wanted the 725th byte you had to read and discard the first 724.

The use of addressing means that any bit can be selected at random, without having to feed out all the preceding bits; hence the name random access. Practically all forms of memory that are used nowadays in microprocessor systems feature random-access addressing, but the name has stuck as a term for read-write memory. Memory of this type needs address pins, data pins and control pins to determine when the chip is enabled and whether it is to be written or read. The reading of a memory of this type does not alter the contents.

A microprocessor must be used in conjunction with memory which is used to hold program and data bytes and also for retaining the results of actions carried out by the microprocessor. Permanent (read-only) memory is needed to hold instructions that need never change (allowing, for example, the microprocessor to be used with a keyboard and a display of some kind). Read-write memory, called RAM, is needed so that it can be loaded with program and data bytes (usually from magnetic disc) that the microprocessor can work on. Only electronic memory operates fast enough to allow the microprocessor to run at its intended speed.

Unlike ROM, RAM is normally volatile – its contents are lost when the supply voltage to the chip is switched off. It is possible, however, to fabricate memory using CMOS techniques and retain the data for very long periods, particularly if a low voltage backup battery can be used. Such CMOS RAM is used extensively in calculators and is used in computers to hold essential setup data, including time and date information.

Static RAM is based on a flip-flop as each storage bit element. The state of a flip-flop can remain unaltered until it is deliberately changed, or until power is switched off, and this made static RAM the first choice for manufacturers in the early days. The snag is that power consumption can be large, because each flip-flop will draw current whether it stores a 0 or a 1. This has led to static RAM being used only for comparatively small memory sizes and where fast operation is needed.

The predominant type of RAM technology for large memory sizes is the dynamic RAM. Each cell in this type of RAM consists of a miniature MOS capacitor with logic 0 represented by a discharged capacitor and logic 1 by a charged capacitor. Since each element can be very small, it is possible to construct very large RAM memory chips (4 M × 1-bit and 16 M × 1-bit are now common), and the power requirements of the capacitor are very small.

The snag is that a small MOS capacitor will not retain charge for much longer than a millisecond, since the connections to the capacitor will inevitably leak. All dynamic memory chips must therefore be refreshed, meaning that each address which contains a logic 1 must be recharged at intervals of

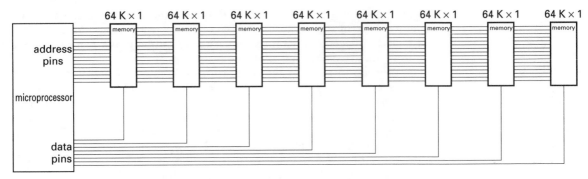

Figure 12.2 How 64 K of memory is arranged for a simple 8-bit computer. Modern machines can use much larger amounts of memory, but the principles are similar.

no more than a millisecond. The refreshing action can be carried out automatically within the chip and the user is never aware of the action.

The capacity of a memory is given in terms of Kbyte or Mbyte (1 Kbyte = 1024 bytes, and 1 Mbyte = 1024 Kbyte). Static RAM is typically used in units of around 250 Kbyte. Dynamic RAM is used for the main memory, typically 8 Mbyte to 128 Mbyte in modern computers.

RAM is volatile, meaning that it loses its data when the power supply is switched off. Static RAM is expensive but very fast, and it is used in small quantities in modern computers. The main memory invariably uses dynamic RAM, mini capacitors which lose data in a short time, a few milliseconds, but which can be refreshed automatically. The low cost of dynamic RAM allows large memories to be used, typically 8 Mbyte upwards.

The buses

The buses of a microprocessor system, as introduced earlier, consist of lines that are connected to each and every part of the system, so that signals are made available at many chips simultaneously and can be passed between any pair of chips. The three main buses are the address bus, the data bus and the control bus. Since understanding the bus action is vitally important to understanding the action of any microprocessor system, we will concentrate on each bus in turn, starting with the address bus.

An address bus consists of the lines that connect between the microprocessor address pins and each of the memory chips in the microprocessor system. In anything but a very simple system, the address bus would connect to other units also, but for the moment we will ignore these other connections. A typical older-style 8-bit microprocessor would use 16 address pins. Using the relationship that n pins allow 2^n binary number combinations, the use of 16 address lines permits 65 536 memory addresses to be used, and modern computing microprocessors use 20, 24, 32 or 64 address lines, of which the use of 32 pins is now most common.

Many types of memory chips are 1-bit types, which allow only 1 bit of data to be stored per address. For a 64 K 8-bit microprocessor, then, the simplest RAM layout would consist of eight 64 K × 1-bit chips, each of which would be connected to all 16 lines of the address bus. Each of these chips would then contribute 1 bit of data, so that each chip is connected to a different line of the data bus. This scheme is illustrated in Figure 12.2.

For modern computers, memory is not installed in single chips. Chips are assembled into units called SIMM (single in-line memory modules) which use a standard plug-in connection (originally using 30 pins but nowadays using 72 pins). Such SIMM units allow memory to be installed in 1 Mbyte to 32 Mbyte chunks.

At each of the 65 536 possible address numbers, each chip will give access to 1 bit, and this access is provided through the lines of the data bus. The combination of address bus and data bus provides for addressing and the flow of data, but another line is needed to determine the direction of data.

This extra line is the read-write line, one of the lines of the control bus (some microprocessors use separate read and write outputs). When the read-write line is at one logic level, the signal at each memory chip enables all connections to the inputs of the memory units, so that the memory is written with whatever bits are present on the data lines. If the read-write signal changes to the opposite logic level, then the internal gating in the memory chips connects to the output of each memory cell rather than to the input, making the logic level of the cell affect the data line. In addition, there are usually one or more enable-disable lines so that the memory can be disabled when addressing is used for other purposes.

Memory chips are connected to the bus lines, and since it is quite common for a chip to store in 1-bit units, one memory chip might be needed for each bit of a full byte. On modern computers, the chips are assembled into SIMM units which can be plugged easily into the computer. In addition to the address and data lines, the memory must use read-write signals to determine the direction of data flow, and enable-disable signals to allow the whole memory to be isolated when the address bus is being used for other purposes.

Note that in practice dynamic memory chips use a rather different addressing system – each address consists of a column number and a row number. This is done to make refreshing simpler, and the address numbers on the address bus need to be changed into this format by a memory manager chip. This does not affect the validity of the description of memory use in this chapter.

The provision of address bus, data bus and read-write lines will therefore be sufficient to allow the older type of microprocessor to work with 64 Kbyte of memory in this example. For smaller amounts of memory, the only change to this scheme is that some of the address lines of the address bus are not used. These unused lines must be the higher-order lines, starting at the most significant line. For a 16-line address bus, the most significant line is designated as A15, the least significant as A0.

A memory system for an 8-bit processor that consisted purely of 64 K of RAM, however, would not be useful, because no program would be present at switch-on to operate the microprocessor. There must be some ROM present, even if it is a comparatively small quantity. For some control applications, the whole of the programming might use only ROM, and the system would consist of one ROM chip connected to all of the data bus lines, and as many of the address lines as were needed to address the chip

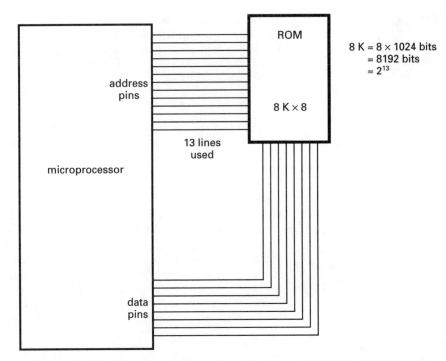

Figure 12.3 Connecting a microprocessor to an 8 K ROM chip.

fully. As an example, Figure 12.3 shows what would be needed in this case, using an 8 K × 8-bit ROM, which needs only the bottom 13 address lines.

It is more realistic to assume that a system will need both ROM and RAM, and we now have to look at how these different sets of memory can be addressed. In the early days, the total addressing capability of an 8-bit machine was no particular restriction, and a common configuration was of 16 K ROM and 16 K RAM. This could be achieved by 'mapping the memory' as shown in Figure 12.4 – other combinations are, of course, possible. In the scheme that is illustrated, the ROM uses the first 16 K of addresses, and the RAM uses the next 16 K. Now the important thing about this scheme is that 16 K corresponds to 14 lines of an address bus, and the same 14 lines are used for both sets of memory.

The lower 14 address lines, A0 to A13, are connected to both sets of chips, represented here by single blocks. Line A14, however, is connected to chip-enable pins, which as the name suggests, enable or disable the chips. During the first 16 K of addresses, line A14 is low, so that ROM is enabled (imagining the enable pin as being active when low) and RAM is disabled. For the next 16 K of addresses on lines A0 to A13, line A14 is high, so that ROM is disabled and RAM is enabled. This allows the same 14 address lines to carry out the addressing of both ROM and RAM. A simple scheme like this is possible only when both ROM and RAM occupy the same amount of memory and require the same number of address lines.

The same address lines are used for both ROM and RAM chips, using enable-disable pins to ensure that one range of address numbers will activate ROM and another range will activate RAM.

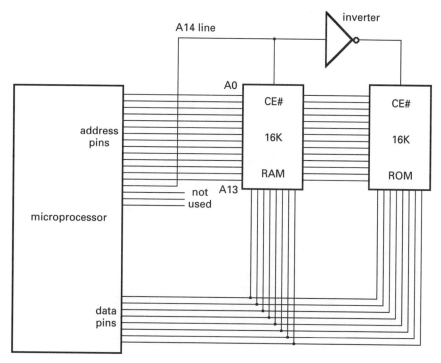

Figure 12.4 Connecting 16 K of RAM and 16 K of ROM so that each memory unit can use different address numbers.

The PC register and addressing

The microprocessor runs a program by outputting address numbers on the address bus so as to select memory. At each memory address, one or more bytes will be read so as to obtain an instruction or the data to carry out an instruction, or bytes can be written to store in memory. The sequence of reading memory is normally a simple incrementing order, so that a program which starts at address 0000 will step to 0001, 0002, and so on, automatically as each part of the program is executed. The exception is in the case of a *jump*, caused by an interrupt (see later) or by a software instruction. A jump in this sense means that a new address will be placed into the program counter register, and the microprocessor will then read a new instruction starting at this address. For the moment, however, the important point is that the normal action is one of incrementing the memory address each time a program action has been executed, and the sequence of actions is all important.

Many microprocessors will read a byte from some fixed address when they start operating, and this byte is a jump instruction, followed by data, that will cause another address range to be used. This allows the programmer to specify where the main program will be located.

The program counter (PC) or instruction pointer (IP) register is the main addressing register, and is connected by gates to the address pins of the microprocessor. The number in this register will be initialised by a voltage applied to a RESET pin, and it is automatically incremented each time an instruction has been executed, or when an instruction calls for another byte.

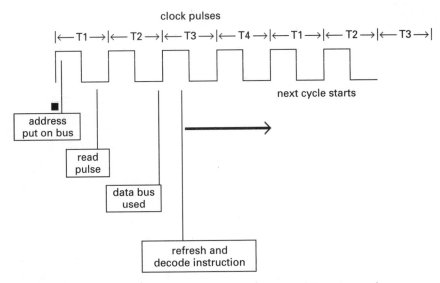

clock pulses

Figure 12.5 Clock pulse and actions. An outline of the stages in carrying out a simple instruction.

Imagine, for example, that the whole of the RAM memory from address 0000 is filled with an NOP instruction byte. NOP means no operation, and its action is simply to do nothing, just go on to the next instruction. If the PC is reset to contain the address 0000, then the NOP byte (which might be 00) at this address will be read, decoded, and acted on. The action is nil, and so the PC is incremented to address 0001, the byte at this address is read, and the action is repeated. Figure 12.5 shows a simplified timing diagram for the Z-80, indicating what actions are involved in reading an instruction byte – note that several clock cycles are needed to carry out one simple instruction.

If the entire memory is filled in this way, the microprocessor will simply cycle through all of the memory addresses until the address reaches 0000 again, and the whole addressing sequence will repeat. Of course, in a real-life system, the memory is not full of NOP bytes. The timing and the PC actions then depend very much on what instruction bytes are present, and even more so on the addressing method.

Looking at addressing methods brings us into the realm of software, but is necessary for understanding how the PC and buses can be used during an instruction that involves the use of memory. The principle is simple enough – many of the instructions of the microprocessor require a byte (or more) to be obtained from the memory.

Carrying out instructions like NOP, or the shift and rotate instruction, do not normally require any other load from memory. This is because these actions are carried out on a single byte, word (16 bits) or Dword (32 bits), which can be stored in one of the registers of the microprocessor. For a lot of actions, though, one word will be stored in a register, and another word must be taken from memory. This is controlled by programming – the instruction byte or word is followed by a set of bytes that is the address number for where the data is stored, or some number that allows the address to be calculated.

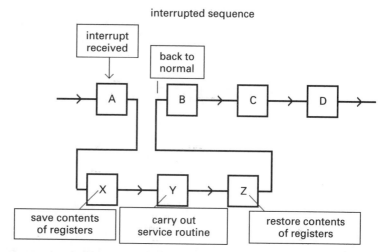

Figure 12.6 How an interrupt works. While the interrupt routine is activated, the state of all registers must be stored in memory so that the microprocessor can recover its register contents when the main program is resumed.

The microprocessor operates in sequence, and when power is applied, it will read a byte from some specified location, which might be the first memory byte but is often arranged to be a byte near the last possible memory address. This first byte will often be a jump instruction that causes the processor to read another byte at another location. From then on, bytes are read in sequence unless a jump command causes a change to another memory location. The program counter (PC) or instruction pointer (IP) register is used to store each address, and unless a jump is encountered, this register is incremented after each instruction.

Interrupts

An interrupt is a signal that interrupts the normal action of the microprocessor and forces it to do something else, almost always a routine which starts at a different address and which will carry out an action that deals with the needs of the interrupt signal. Such a routine is called an *interrupt service routine*.

The interrupt is an electrical signal which is applied to one of the interrupt pins of the microprocessor – it can come from a switch or by way of a software instruction. When this signal is received, the microprocessor executes an interrupt routine. In doing so, it will complete the instruction that it is processing, and then jump to an address to get directions for a service routine, in this example the routine that reads the keyboard.

When the service routine is completed, the microprocessor resumes its program instructions, Figure 12.6. All microprocessors allow a part of the memory to be designated as a *stack*. This means simply that some addresses are used by the microprocessor for storing register contents, making use of the memory in a very simple last-in-first-out way.

Precisely which addresses are used in this way is generally a choice for the software designer. When an interrupt is received, the first part of the action is for the microprocessor to complete the action on which it is engaged. The next item is to store the PC address in the stack memory. This action is completely automatic. Only the address is stored in this way,

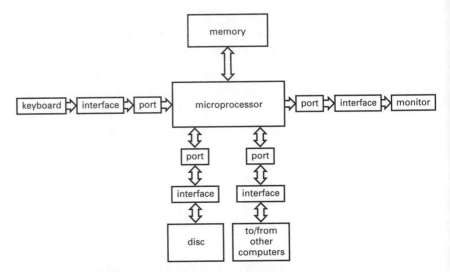

Figure 12.7 A block diagram for a computer, omitting details.

however. If the interrupt service routine will change the contents of any other registers of the microprocessor, it will also be necessary to save the contents of these registers on the stack.

This is something that has to be attended to by the programmer who writes the interrupt service routine, saving the register contents at the start of the interrupt service routine and replacing them afterwards. Finally, the problem of what happens if an interrupt signal arrives while another interrupt is being serviced is dealt with by disabling the interrupt mechanism while an interrupt is being serviced. This is not necessarily automatic, and will often form one of the first items in the interrupt service routine. The tendency over the short history of microprocessors has been to delegate these actions to the software writer rather than to embed them in hardware.

The automatic action of reading bytes in succession can be interrupted by a signal to an interrupt pin. This action will cause the microprocessor to start taking instructions (the service routine) from another location in the memory, and when this program (a subroutine) is completed, the previous action will be restored. Essential data is stored in part of the memory, called the stack, when an interrupt is called, and is restored when the interrupt is completed.

Inputs and outputs A system which consisted of nothing more than a microprocessor and memory could not do much more than mumble to itself. Every useful system must have some method of passing bytes out of the system to external devices, and also into the system from external devices. For computer use, this means at the very least the use of a VDU screen and a keyboard.

For modern systems, you can add to this disc drives, floppy, hard or CD-ROM, various sockets where other add-ons can be plugged (such as printer, light pen, joystick, a mouse, trackerball, etc.), and possibly A-to-D inputs for measuring voltages. If the microprocessor system is intended for machine-control purposes, then the outputs will be very important indeed, because these will be the signals that will control the machine. Figure 12.7 shows a block diagram for a computer system using a microprocessor.

The inputs may be instruction codes from a disc, pulses from limit switches, digital voltage readings from measuring instruments, and so on. Whatever the function of the system, then, the inputs and outputs form a very important part of it all. A surprising number of control actions of a microprocessor system, in fact, consists of little more than passing an input signal to an output device, perhaps with some monitoring or comparison action thrown in.

Before we get involved with the details of how signals are passed between a microprocessor system and the rest of the world, we need to be aware of the problems that are involved. The main problem is one of timing. The microprocessor system works fast, governed by the rate of the system clock. There is no point in having instructions in the software that will make the microprocessor send a word out of the system unless you can be sure that whatever you are sending the word to can deal with it at that time.

The time that is involved might be perhaps a couple of clock cycles, a fraction of a microsecond, and there are not many systems, apart from another microprocessor, that can deal with such short-duration signals. The same problem applies to incoming signals. At the instant when you press a key on a keyboard, can you be certain that a microprocessor is executing an instruction that will read the key? Many such problems are dealt with by using the interrupt system, so that pressing a key on the keyboard will generate an interrupt that will activate a reading routine.

Another matter is the nature of the signals. The signals from a microprocessor system will be digital signals, using standard voltages of 0 V and +5 V. If the input from another system happens to be an analogue signal of maximum amplitude 50 mV, or if the signal that is required by a device at the output is a +50 V, 500 Hz AC signal, then a lot of thought will have to be devoted to interfacing. This is outside the scope of this chapter, because for machine-control applications in particular, interfacing is by far the most difficult action to achieve in a new design.

Summary

A microprocessor and its memory becomes a computer when it is connected to a keyboard, display unit, disc drive(s) and power supply. The signals may have to be converted to other formats, or their timing altered, to be usable by these other units.

Ports

A port is a circuit which controls the transfer of signals into or out of the microprocessor system. If the only requirement for a system is to pass a single bit in and out, the obvious method is to use a simple latch chip, connecting with one of the data lines of the microprocessor system. Even if eight data lines are needed, a straightforward hardware latch may be sufficient, particularly if the signals are all in one direction, or if most of the signals are in one direction. Ports must be used, for example, for your keyboard inputs.

The port will have an address, like memory, and it can store a small amount of data, perhaps as much as four bytes. To pass data out of the system, the microprocessor uses the port address and puts data on the data bus. The port stores the data, and sends out a signal that it contains data. The circuits beyond the port can then read the signal as and when they are ready.

In the other direction, if an external circuit sends data to the port, the port will store the data and send an interrupt signal to the microprocessor. The interrupt service routine for the port will start running, allowing the microprocessor to address the port and read the data. When this has been

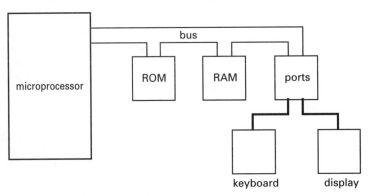

Figure 12.8 A block diagram that is typical of any simple calculator. All units apart from keyboard and display are usually made as a single chip.

done, the port will send out a signal to indicate that it is ready for more data, and the microprocessor resumes its normal actions.

Port chips are almost as complicated as microprocessor chips, and are used to connect the microprocessor buses to the external components of a computer system. Ports are programmable chips, and they carry out all the interfacing actions that are needed to make the signals compatible.

Calculators

A calculator is constructed using a microprocessor and memory, with most of the memory in ROM form to provide the routines for the more complicated processes. The arithmetic actions of the microprocessor are confined to addition or subtraction of two bytes, sometimes with provision for multiplication and division of single bytes. Anything more complicated must be handled by software – program instructions – and these can be read from the ROM.

Virtually all calculator actions require these additional program instructions, because the calculator uses ordinary denary numbers that can use a decimal point, but the microprocessor will handle single bytes that represent whole numbers. The programs in the ROM are therefore the 'clever' part of a calculator, and the rest is well standardised, a microprocessor with a small amount of RAM (for storing answers, intermediate results and constants) and an I/O port for the keyboard and the display, Figure 12.8. A few calculators are programmable, meaning that the instructions that you would normally carry out in sequence on data can be programmed into RAM and carried out on data that you can also store in RAM.

The diagram of Figure 12.8 shows separate units, but the whole of a calculator, including microprocessor, ROM, RAM, and ports, is usually constructed as a single chip, needing only connections to display, keyboard, switches and battery.

Computers

Figure 12.9 shows a block diagram for a generalised 8-bit data computer system which would have been typical of small computers in the early 1980s, and the principles of which can still be applied to modern 32-bit

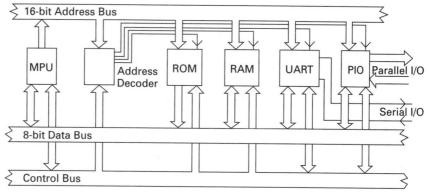

Figure 12.9 A block diagram for an 8-bit computer, typical of types produced in the mid-1980s. Modern machines are more elaborate, but the block diagram is typical of basic design.

machines. The block that is labelled as MPU is the microprocessor unit which will have its own internal buses.

The MPU is shown as placing signals on the address bus, because in a system like this with only one processor, the MPU has total control over the address bus – all other chips *receive* address signals. The data connection between the data bus and the MPU is two-way, because the MPU must read data as an input and also output processed data. The connection of the MPU to the control bus is also two-way, because the MPU will issue some control signals to other chips and also receive signals from other chips. Note that ROM sends data to the data bus but does not receive data.

The address *decoder* is the chip that allocates addresses. Some MPU address numbers must correspond to storage in ROM and some to storage in RAM, and a few are reserved for the UART (serial) and PIO (parallel) input/output chips. The address decoder chip receives signals from the address bus and from the control bus (originating in the MPU) and determines which of the four memory-using sections will be enabled. This chip prevents any possibility of contention for the use of the buses, because only the chip that is entitled to use the buses will be enabled.

The *ROM* unit (usually one or two chips) receives signals from the address bus, and is enabled or disabled by an output from the address decoder. It is also controlled by control bus signals, and its output is placed on the data bus. This output will for the most part be program instructions to the MPU, such as start-up instructions, and short routines for such tasks as peripheral control (disc, VDU, keyboard etc.).

The *RAM* unit (which might consist of a large number of individual chips, each storing one bit of data) uses the signals from the address bus and is enabled by the address decoder. It also receives signals from the control bus, notably the read-write signals that determine the direction of data flow, and it has a bidirectional (two-way) connection with the data bus.

The *UART* chip uses the address bus signals and an enabling signal from the address decoder. It has a bidirectional connection with the data bus and also with the control bus. The control bus can determine the direction of data and can also interrupt the MPU to ensure that serial input is dealt with by a suitable routine. Two separate lines allow for external connections to RS-232 connectors for serial input and output.

The *PIO* chip uses the same scheme of address bus connection and enabling line from the address decoder. It has a bidirectional connection with the data bus and also with the control bus, so that it also can send signals to the MPU to interrupt it and run an input routine if required. This chip also has the parallel I/O connections to the parallel port connector. Though the PIO is usually operated with a printer as an output only, it can be used by other units bidirectionally, making it possible to connect scanners and disc drive units through the parallel connector.

All of the computer actions are determined by the software, and a computer without software is no more useful than a record-player without records. A small amount of software is permanently fixed into the ROM of the computer system, and this can be used to allow a few keyboard actions and, even more important, to allow more software to be read into the memory from a magnetic disc, the *hard drive*. This software is the operating system, which consists of a set of routines for using other programs, allowing all the effects that we take for granted. Software, and the details of computer systems, are outside the scope of this book but if you are interested, take a look at some of the books in the Made Simple series (Butterworth-Heinemann) that are devoted to computing.

Microprocessors and computing methods are used in many applications that are not quite so obvious. The engine management system of a modern car is a microprocessor-based computer system that takes inputs such as engine speed, air temperature, throttle opening and so on to control fuel injection and ignition timing. Even the humble washing machine is now likely to use a microprocessor system to replace the old methods based on electric clock mechanisms with cams and switches.

Calculators and computers use a microprocessor as the main programmable unit. Calculators generally use only a number keyboard and a number display, with a small memory. Computers use a much wider range of connections such as hard and floppy drives, CD units, sound boards, scanners, light pens, full-sized keyboards and CRT monitors.

Chapter 13

Miscellaneous systems

This chapter is concerned with some items that were omitted earlier, either because they did not match the content of other chapters or because you would have had some difficulties in understanding them until other chapters had been read. Much of the chapter is devoted to a more detailed description of compact disc recording, because this is by far the most advanced electronics system that has become a familiar consumer item, and the methods that are used for CD are applied to a wide range of other digital systems.

The oscilloscope

This description assumes that you have read the description of the cathode-ray tube in Chapter 1, and it refers to the older analogue type of oscilloscope. The block diagram of the electronic circuits of a simple oscilloscope, omitting the CRT, power supplies and brightness controls, is shown in Figure 13.1. A signal applied to the Y input is attenuated by a calibrated variable attenuator (usually a switch rather than a potentiometer) before being applied to the Y amplifier. The amplified output drives the Y plates, which cause the spot to be deflected in the vertical direction. The gain of the amplifier and the settings of the attenuator are matched to the sensitivity of the CRT, and the attenuator is consequently calibrated in volts/cm.

Some of the Y input signal is fed to a synchronising circuit which generates a pulse at every signal peak (either positive or negative peaks, usually selected by a switch). Each of these synchronising pulses is then used to start a timebase, a sawtooth signal which is applied by the X amplifier to the X plates.

The effect of such a sawtooth waveform is to deflect the electron beam, and so the spot, at a steady speed across the screen in what is called a sweep, and then to return it very rapidly indeed to its starting point in what is called flyback. The speed of the timebase sweep can be controlled by a calibrated switch, so that the time needed for the spot to scan every centimetre of the screen can be printed on the switch plate.

Another switch enables the X amplifier to be used to handle other signals, if desired. When the CRT is in use, a signal of unknown frequency and amplitude is applied to the Y input. The Y attenuator and the timebase speed controls should then be adjusted until a steady display of measurable size appears on the screen. The peak-to-peak amplitude of the waveform is then found by measuring, with the aid of the calibration of the attenuator, the vertical distance between peaks on the graticule, a transparent calibrated sheet covering the screen. The time-period of one cycle (τ) is measured on the same graticule, using this time the calibration of the timebase. The frequency of the signal can then be calculated by the formula: $f = 1/\tau$.

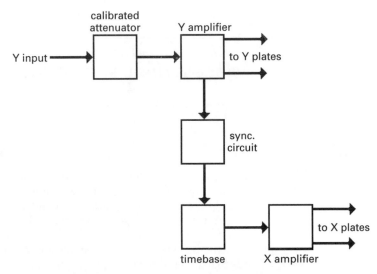

Figure 13.1 Principles of the oscilloscope, the essential instrument for working with waveforms.

The oscilloscope is used for electronics measurements, particularly for waveform measurements on analogue circuits. The principle is that a CRT beam is deflected horizontally by a sawtooth waveform, the timebase, while the vertical deflection is caused by an input signal. If the timebase speed is correctly matched to the frequency of the incoming waveform, one or more cycles of the wave will be displayed, and a measuring graticule allows voltage and time measurements to be carried out.

Modems

The word modem is derived from the words modulation and demodulation, and is used for a device that allows computer data to be passed along telephone lines or by radio links. A telephone line was never intended for more than voice signals, so that it operates best with frequencies in the range of 300 to 400 Hz. By contrast, computer signals consist of successive 0 and 1 voltage levels which may be at a speed of many MHz. Clearly it is impossible to transfer unaltered computer signals along an ordinary telephone cable. To achieve communication of data along telephone lines the data rate must be reduced, and some form of modulation must be used. This is the basis of the modulation action of the modem, and a block diagram of how the modem is used to link computers is illustrated in Figure 13.2.

Any sine wave carrier can be modulated in three different ways, by

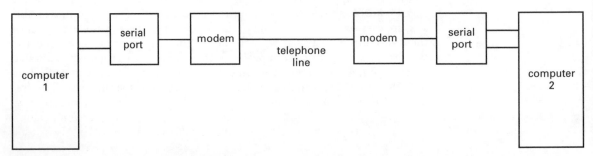

Figure 13.2 Using modems to allow computers to communicate by telephone lines. This is the basis of electronic mail and the Internet.

modulating its amplitude, its frequency or its phase. This gives rise to the different methods that are used in modems, called amplitude shift keying, frequency shift keying, and phase shift keying.

The *amplitude shift keying* (ASK) system alters the amplitude of the carrier according to the level of the data signal. At its simplest, the carrier can simply be switched on for logic 1 and off for logic 0 (this is also known as on-off-keying or OOK). This type of modulation has fallen out of use because of its poor S/N ratio and therefore its high bit error rates. The problem is that ASK signals and noise signals are very similar, so that a receiver cannot distinguish genuine signals from noise signals.

A *frequency shift keying* (FSK) system involves switching the carrier wave between two frequencies. In the 1970s, small computers used one form of this, known as Kansas City (or CUTS) modulation, for tape recording digital signals, and this form is a good example of the method. For the CUTS standard bursts of eight cycles of 2400 Hz or four cycles of 1200 Hz are used to represent 1 or 0 respectively. FSK systems for modems have used other frequency values, often with a frequency ratio that is not exactly 2:1. FSK has a better bit error rate than ASK under the same conditions, but requires a wider transmission bandwidth.

Phase shift keying (PSK) is a method that is now the most commonly used for low-speed modems. This uses a single carrier frequency where the phase is altered by the data signal. For example, a logic 0 might cause no phase change, but a logic 1 will shift the carrier phase by 180°. Of the three methods, PSK has by far the best lowest bit error rate and the bandwidth that it requires is least.

A more modern development of PSK is *Multi-phase PSK*, in which the carrier phase can be shifted to several different values. For example, if eight different values of phase can be used, then each digital signal represented by a phase change can consist of three bits (since a 3-bit signal can convey number values of 0 to 7, eight in all). If phase and amplitude modulation are combined, it is possible to carry four bits of digital data per unit carrier. Such systems are also used for the NICAM stereo sound system, see later.

This allows a carrier of, say, 350 Hz, to convey much more information than might be expected, and bit rates of 9600 per second or more are normal. The effective bit rate can be increased by using data compression (eliminating redundant bits in signals) and by using faster rates along with error detecting and correcting methods. Modems are now used to link computers that can be located anywhere in the world, and pass information over any distance.

In practice, a modem is connected to the serial port of the computer, and software is used to set up the modem and the port for the required data rate, the telephone number to use, etc. When a communication is to be sent, the software will make the modem open the line and dial the number, and when the remote computer answers, the two confirm connection and the data is sent. At the other end, the modem has responded to the ringing tone and has opened the line. The signals from the transmitting modem are analysed, the receiving modem automatically sets itself to use the same system. It then acknowledges contact, and receives the signals, which are usually stored as a file on the disc.

Summary

The modem (modulator-demodulator) is used to connect computers over telephone lines. A low-frequency carrier is modulated by the digital signals from a low-speed serial port. By using suitable modulation systems, rates of 9600 bits per second and higher can be achieved with a very low error rate. Modern software allows the modem to be controlled very easily, and enables automatic dialling and answering.

CD system

The CD system is one rare example of international co-operation leading to a standard that has been adopted worldwide, something that was notably lacking in the first generation of videocassette recorders. Whereas when videocassette recorders appeared, they performed an action that was not possible before, audio compact discs were in competition with existing methods of sound reproduction, and it would have been ridiculous to offer several competing standards.

In 1978, a convention dealing with digital disc recording was organised by some 35 major Japanese manufacturers which recommended that development work on digital discs should be channelled into 12 directions, one of which was the type proposed by Philips. The outstanding points of the Philips proposal were that the disc should use constant linear velocity recording, 8-to-14 modulation, and a new system of error correction called Cross-interleave Reed–Solomon code (CIRC).

Constant velocity recording means that no matter whether the inner or the outer tracks are being scanned, the rate of digital information should be the same, and Philips proposed to do this by varying the speed of the disc at different distances from the centre so that the rate of reading was constant. You can see the effect of this for yourself if you play a track from the start of a CD (the inside tracks) and then switch to a track at the end (outside). The reduction in rotation speed when you make the change is very noticeable, and this use of constant digital rate makes for much simpler processing.

By 1980, Sony and Philips had decided to pool their respective expertise, using the disc modulation methods that had been developed by Philips along signal-processing systems developed by Sony. The use of a new error-correcting system along with higher packing density made this system so superior to all other proposals that companies flocked to take out licences for the system. All opposing schemes died off, and the CD system emerged as a rare example of what can be achieved by co-operation in what is normally an intensely competitive market.

The audio CD system is a global standard, obtained by pooling the expertise of Philips and Sony. From the start, the system was designed to produce much higher performance than was obtainable using analogue methods, and provision was made for other uses, allowing CD to be used in later developments such as CD-ROM for computers.

The optical system

The CD system makes use of optical recording, using a beam of light from a miniature semiconductor laser. Such a beam is of low power, a matter of milliwatts, but the focus of the beam can be to a very small point, about 0.6 μm in diameter – for comparison, a human hair is around 50 μm in diameter. The beam can be used to form pits in a flat surface, using a depth which is also very small, of the order of 0.1 μm. If no beam strikes the disc, then no pit is formed, so that we have here a system that can digitally code pulses into the form of pit or no-pit. These pits on the master disc are converted to pits of the same scale on the copies. The pits/dimples are of such a small size that the tracks of the CD can be much closer – about 60 CD tracks take up the same width as one LP track.

Reading a set of pits on a disc also makes use of a semiconductor laser, but of much lower power since it need not vaporise material. The reading beam will be reflected from the disc where no pit exists, but scattered where

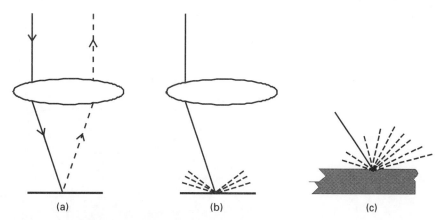

Figure 13.3 The compact disc. Light reaching a flat portion is reflected back (a), but light striking a pit is scattered so that there is no strong reflected beam (b). A close-up is shown in (c).

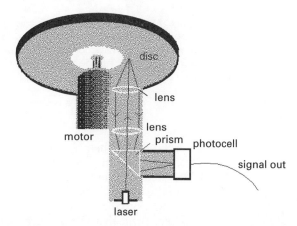

Figure 13.4 The optical arrangement of a CD player. The motor spins the disc and the laser light is focused to a spot which is reflected where there is no pit on the disc. The reflected light is diverted by the prism and falls on the photocell, causing an output.

there is a pit, Figure 13.3. By using an optical system that allows the light to travel in both directions to and from the disc surface (Figure 13.4), it is possible to focus a reflected beam on to a photodiode, and pick up a signal when the beam is reflected from the disc, with no signal when the beam falls on to a pit. The output from this diode is the digital signal that will be amplified and then processed eventually into an audio signal. Only light from a laser source can fulfil the requirements of being perfectly *monochromatic* (one single frequency) and *coherent* (no breaks in the wave-train) so as to permit focusing to such a fine spot.

The CD player uses a beam that is focused at quite a large angle, and with a transparent coating over the disc surface which also focuses the beam as well as protecting the recorded pits. Though the diameter of the beam at the pit is around 0.5 μm, the beam diameter at the surface of the disc, the transparent coating, is about 1 mm. This means that dust particles and hairs on the surface of the disc have very little effect on the beam, which passes

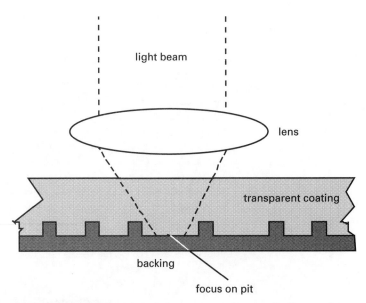

light beam

lens

transparent coating

backing

focus on pit

Figure 13.5 Illustrating that the beam has a larger radius as it passes through the transparent coating. This avoids errors due to surface scratches. The ratio of beam diameter at the transparent surface to beam diameter at the reflective surface is enormous – about 2000 times.

on each side of them – unless your dust particles are a millimetre across! This, illustrated in Figure 13.5, is just one of the several ways in which the CD system establishes its very considerable immunity to dust and scratching on the disc surface, the other being the advanced bit-error detection and correction system.

Given, then, that the basic record and replay system consists of using a finely focused beam of laser light, how are the pits arranged on the disc and how does the very small reading spot keep track of the pits? To start with the pits are arranged in a spiral track, like the groove of a conventional record. This track, however, starts at the *inside* of the disc and spirals its way to the outside, with a distance between adjacent tracks of only 1.6 μm. Since there is no mechanical contact of any kind, the tracking must be carried out by a servo-motor system that guides the laser and lens assembly. The principle is that the returned light will be displaced if there is mistracking, and the detector unit contains two other photodiodes to detect signals that arise from mistracking in either direction. These signals are obtained by splitting the main laser beam to form two side-beams, neither of which should ever be deflected by the pits on the disc when the tracking is correct. These side beams are directed to the photodiodes. The signals from these diodes are used to control the servo-motors in a feedback loop system, so that the corrections are being made continually while the disc is being played. In addition, the innermost track carries the position of the various music tracks in the usual digitally coded form. Moving the scanner system radially across the tracks will result in signals from the photodiode unit, and with these signals directed to a counter unit, the number of tracks from the inner starting track can be counted. This therefore allows the scanner unit to be placed on any one of the possible 41 250 tracks.

A servo motor is a small motor that is used in a feedback loop. Electronic sensors detect that there is an error in position, and operate the servo-motor until the error is zero, at which point there is no output to the servo motor.

Positioning on to a track is a comparatively slow operation because part of each track will be read on the way before the count number is increased. In addition, the motor that spins the disc is also servo-controlled so that the speed of reading the pits is constant. Since the inner track has a diameter of 50 mm and the maximum allowable outer track can have a diameter of 116 mm, the number of pits per track will also vary in this same proportion, about 2.32. The disc rotation at the outer edge is therefore slower by this same factor when compared to the rotational speed at the inner track. The rotational speeds range from about 200 rpm on the inner track to about 500 rpm on the outer track – the actual values depend on the servo control settings and do not have to be absolute in the way that the old $33\frac{1}{3}$ speed for vinyl discs had to be absolute. The only criterion of reading speed on a CD is that the pits are being read at a correct rate, and that rate is determined by a master clock frequency. This corresponds to a reading speed of about 1.2 to 1.4 metres of track per second.

The next problem is of recording and replaying two channels for a stereo system, because the type of reading and writing system that has been described does not exactly lend itself well to twin-channel use with two tracks being read by two independent scanners. This mechanical and optical impossibility is avoided by recording the two-channel information on a single track, by recording samples from the channels alternately. This is made possible by the storage of data while it is being converted into 'frame' form for recording, because if you have one memory unit containing L-channel data and one containing R-channel data, they can be read alternately into a third memory ready for making up a complete frame. There is no need to have the samples taken in a phased way to preserve a time interval between the samples on different channels. As the computer jargon puts it, the alternation does not need to take place in *real time*.

The optical system that is used for recording or reading CDs is based on low-power semiconductor lasers where the beam can be focused to a very small point. On recording, each logic 1 is reproduced as a pit on the finished CD. For reading, each pit scatters the laser beam so that it is not reflected back to a detector. The guidance system uses two other detectors which produce a constant output while the main beam is correctly aimed. This is achieved using a servo-motor system with the error signals used to correct the position, and the same system is used to move from one part of the continuous track to another. Stereo recording is achieved by recording L and R data alternately, using storage so that the signals can be reassembled correctly.

Digital processing

We now have to deal with the most difficult parts of the whole CD system. These are the modulation method, the error-detection and correction, and the way that the signals are assembled into frames for recording and replay. The exact details of each of these processes are released only to licensees that buy into the system. This is done on a flat-rate basis – the money is shared out between Philips and Sony, and the licensee receives a manual called the 'Red Book' which precisely defines the standards and the meth-

(a) Data 12×16 bits $= 24 \times 8$ bits ⎤

     Error correction 4×16 bits $= 8 \times 8$ bits ⎬ 33×8 bits

     Control/display $= 1 \times 8$ bits ⎦

                                      264 bits per frame

(b) Total data of 33 bytes 33×14 bits $= 462$ bits

     Synchronisation $= 24$ bits

     Redundant bits 3×34 $= 102$ bits

                                        588 bits per frame

Figure 13.6 How the signals are organised into a frame of data (a) as bytes, (b) after 8-to-14 modulation.

ods of maintaining them. Part of the agreement, as you might expect, is that the confidentiality of the information is maintained, so that what follows is an outline only of methods. This is all that is needed, however, unless you intend to design and manufacture CD equipment, because for servicing work you do not need to know the circuitry of an IC or the content of a memory in order to check that it is working correctly.

The error-detection and correction system is called CIRC – Cross-Interleave Reed–Solomon Code. This code is particularly suitable for the correction of what are termed 'long burst' errors, meaning the type of error in which a considerable amount of the signal has been corrupted. This is the type of error that can be caused by scratches, and the code used for CDs can handle errors of up to 4000 bits, corresponding to a 2.5 mm fault on the disc lying along the track length. At the same time, a coding that required too many redundant bits to be transmitted would cramp the expansion possibilities of the system, so the CIRC uses only one additional bit added to each three of data – this is described as an efficiency of 75%. Small errors can be detected and remedied, and large errors are dealt with by synthesising the probable waveform on the basis of what is available in the region of the error.

The whole system depends heavily on a block or *frame* structure which, though not of the same form as a TV signal frame, allows for the signal to carry its own synchronisation pattern. This makes the recording self-clocking, so that all the information needed to read the signal correctly is contained on the disc – there is no need, for example, for each player to maintain a crystal-controlled clock pulse generator working at a fixed frequency. In addition, the presence of a frame structure allows the signal to be recorded and replayed by a rotating head type of tape recorder, so that the digital signals from a master recorder can be transferred to a CD without the need to convert to analogue and back.

Each frame of data contains a synchronisation pattern of 24 bits, along with 12 units of data (using 16-bit 'words' of data), four error-correcting words (each 16 bits) and 8 bits of control and display code. Of the 12 data words, six will be left-channel and six right-channel, and the use of a set of 12 in the frame allows expansion to four channels (three words each) if needed later. Though the actual data words are of 16 bits each, these are split into 8-bit units for the purposes of assembly into a frame. The content of the frames, before modulation and excluding synchronisation, is therefore as shown in Figure 13.6(a).

The digital signals are assembled into a frame, along with synchronisation bits, error-correcting words and control codes, amounting to 264 bits per frame. This is the basic digital unit, and it is then modulated into its final format.

EFM modulation

The word 'modulation' is used here to mean the method that is used to code the 1s and 0s of a digital number into 1s and 0s for recording purposes. The principle, as mentioned earlier, is to use a modulation system that will prevent long runs of either 1s or 0s from appearing in the final set of bits that will burn in the pits on the disc, or be read from a recorded disc, and calls for each digital number to be encoded in a way that is quite unlike binary code. The system that has been chosen is EFM, meaning 8-to-14 modulation, in which each set of eight bits is coded on to a set of 14 bits for the purposes of recording. The number carried by the eight bits is coded as a pattern on the 14 bits instead, and there is no straightforward mathematical relationship between the 14-bit coded version and the original 8-bit number.

The purpose of EFM is to ensure that each set of eight bits can be written and read with minimum possibility of error. The code is arranged, for example, so that changes of signal never take place closer than three recorded bits apart. This cannot be ensured if the unchanged 8- or 16-bit signals are used, and the purpose is to minimise errors caused by the size of the beam which might overlap two pits and read them as one. The 3-bit minimum greatly eases the requirements for perfect tracking and focus. At the same time, the code allows no more than 11 bits between changes, so avoiding the problems of having long runs of 1s or 0s. In addition, three redundant bits are added to each 14, and these can be made 0 or 1 so as to break up long strings of 0s or 1s that might exist when two blocks of 14 bits were placed together. The conversion can be carried out by a small piece of fixed memory (ROM) in which using an 8-bit number as an address will give data output on fourteen lines. The receiver will use this in reverse, feeding fourteen lines of address into a ROM to get eight bits of number out.

The use of EFM makes the frame considerably larger than you would expect from its content of 33 8-bit units. For each of these units we now have to substitute 14 bits of EFM signal, so that we have 33 × 14 bits of signal. In addition, there will be three additional bits for each 14-bit set, and we also have to add the 24 bits of synchronisation and another three redundant bits to break up any pattern of excessive 1s or 0s here. The result is that a complete frame as recorded needs 588 bits, as detailed in Figure 13.6(b). All of this, remember, is the coded version of 12 words, six per channel, corresponding to six samples taken at 44.1 kHz, and so representing about 136 μs of signal in each channel.

Normal binary signals, as assembled into a frame, are liable to contain rapid alternations (like 1010101 . . .) and also long sequences of the same bit type, and neither of these is acceptable for recording purposes. By expanding each 8-bit unit to 14 bits, the code can be changed so that it avoids rapid alternations and long sequences. This 8-to-14 code then has other bits added for separation and synchronisation to make a complete frame occupy 588 bits for 12 words of data.

Error correction

The EFM system is by itself a considerable safeguard against error, but the CD system needs more than this to allow, as we have seen earlier, for scratches on the disc surface that cause long error sequences. The main error correction is therefore done by the CIRC coding and decoding, and one strand of this system is the principle of *interleaving*. Before we try to unravel what goes on in the CIRC system, we need to look at interleaving and why it is carried out.

As we have noted earlier, the errors in a digital recording and replay system can be of two main types, random errors and burst errors. Random errors, as the name suggests, are errors in a few bits scattered around the disc at random and, because they are random, they can be dealt with by relatively simple methods. The randomness also implies that in a frame of 588 bits, a bit that was in error might not be a data bit, and the error could in any case be corrected reasonably easily. Even if it were not, the use of EFM means that an error in one bit does not have a serious effect on the data.

A block error is quite a different beast, and is an error that involves a large number of consecutive bits. Such an error could be caused by a bad scratch in a disc or a major dropout on tape, and its correction is very much more difficult than the correction of a random error. Now if the bits that make up a set were not actually placed in sequence on a disc, then block errors would have much less effect. If, for example, a set of 24 data units (bytes) of 8 bits each that belonged together were actually recorded so as to be on eight different frames, then all but very large block errors would have much the same effect as a random error, affecting only one or two of the byte units. This type of shifting is called *interleaving*, and it is the most effective way of dealing with large block errors. The error detection and correction stages are placed between the channel alternation stage and the step at which control and display signals are added prior to EFM encoding.

The CIRC method uses the Reed–Solomon system of parity coding along with interleaving to make the recorded code of a rather different form and different sequence as compared to the original code. Two Reed–Solomon coders are used, each of which adds four parity 8-bit units to the code for a number of 8-bit units. The parity system that is used is very complicated, unlike simple single-bit parity, and it allows an error to be located and signalled. The CD system uses two different Reed–Solomon stages, one dealing with 24 bytes (8-bit units) and the other dealing with 28 bytes (the data bytes plus parity bytes from the first one), so that one frame is processed at a time. In addition, by placing time delays in the form of serial registers between the coders, the interleaving of bytes from one frame to another can be achieved. The Reed–Solomon coding leaves the signal consisting of blocks that consist of correction code (1), data (1), data (2) and correction code (2), and these four parts are interleaved. For example, a recorded 32-bit signal might consist of the first correction code from one block, the first data byte of the adjacent block, the second data byte of the fourth block, and the second correction code from the eighth data block. These are assembled together, and a cyclic redundancy check number can be added.

At the decoder, the whole sequence is performed in reverse. This time, however, there may be errors present. We can detect early on whether the recorded 'scrambled' blocks contain errors and these can be corrected as far as possible. When the correct parts of a block (error code, data, data, error code) have been put together, then each word can be checked for errors, and these corrected as far as possible. Finally, the data is stripped of all added codes, and will either be error-free or will have caused the activation of some method (like interpolation) of correcting or concealing gross errors.

Further error protection is achieved by interleafing frames, so that an error is less likely to affect a set of consecutive frames. The Reed–Solomon system provides a very considerable amount of detection and correction ability, and if all else fails, the signal will be interpolated.

Production methods

A compact disc starts as a glass plate which is ground and polished to 'optical' flatness – meaning that the surface contains no deformities that can be detected by a light beam. If a beam of laser light is used to examine a plate like this, the effect of reflected light from the glass will be to add to or subtract from the incident light, forming an *interference pattern* of bright and dark rings. If these rings are perfectly circular then the glass plate is perfectly flat so that this can be the basis of an inspection system that can be automated.

The glass plate is then coated with *photo-resist*, a material which hardens on exposure to light (like the old gum-bichromate photographic process). Various types of photo-resist have been developed for the production of ICs, and these are capable of being printed with very much finer detail than is possible using the older type of photo-resist which is used for printed circuit boards, and the thickness and uniformity of composition of the resist must both be very carefully controlled.

The image is then produced by treating the glass plate as a compact disc and writing the digital information on to the photo-resist with a laser beam. Once the photo-resist has been processed, the pattern of pits will develop and this comprises the glass master. The surface of this disc is then silvered so that the pits are protected, and a thicker layer of nickel is then plated over the surface. This layer can be peeled off the glass and is the first metal master. The metal master is used to make ten mother plates, each of which will be used to prepare the stamper plates.

Once the stamper plates have been prepared, mass production of the CDs can start. The familiar plastic discs are made by injection moulding (the word 'stamper' is taken from the corresponding stage in the production of black vinyl discs) and the recorded surface is coated with aluminium, using vacuum vaporisation. Following this, the aluminium is protected by a transparent plastic which forms part of the optical path for the reader and which can support the disc label. The system is illustrated in outline in Figure 13.7.

The CDs are produced by using photo-resist material on a flat glass plate. This is then recorded using a laser that will harden the photo-resist so that after processing a master can be produced. This is used to produce plates which will stamp out plastic copies.

Control bytes

For each block of data one control byte (8 bits) is added. This allows eight channels of additional information to be added, known as sub-codes P to W. This corresponds to one bit for each channel in each byte. To date, only the channels P and Q have been used. The P channel carries a selector bit which is 0 during music and lead-in, but is set to 1 at the start of a piece, allowing a simple but slow form of selection to be used. This bit is also used to indicate the end of the disc, when it is switched between 0 and 1 at a rate of 2 Hz in the lead-out track.

The Q channel contains more information, including track number and timing. A channel word consists of a total of 98 bits, and since there is one

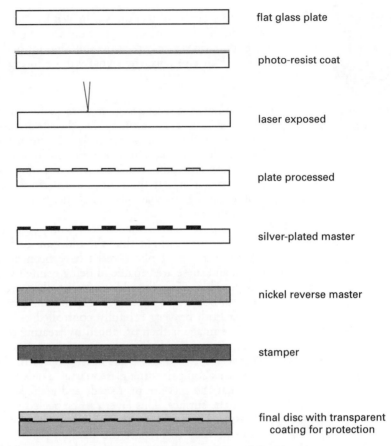

flat glass plate

photo-resist coat

laser exposed

plate processed

silver-plated master

nickel reverse master

stamper

final disc with transparent
coating for protection

Figure 13.7 Production of CDs from glass-plate stage.

bit in each control byte for a particular channel, the complete channel word is read in each 98 blocks. This word includes codes which can distinguish between four different uses of the audio signals.

The allocation of these control channels allows considerable flexibility in the development of the CD format, so that players in the future could make use of features which have to date not been thought of. Some of the additional channels are used on the CD-ROMs which are employed as large-capacity storage for computers.

The end-result

All of this encoding and decoding is possible mainly because we can work with stored signals, and such intensive manipulation is tolerable only because the signals are in the form of digital code. Certainly any efforts to correct analogue signals by such elaborate methods would not be welcome and would hardly add to the fidelity of reproduction. Because we are dealing with signals that consist only of 1s and 0s, however, and which do not provide an analogue signal until they are converted, the amount of work that is done on the signals is irrelevant. This is the hardest point to accept to anyone who has been brought up in the school of thought that the less that is done to an audio signal the better it is. The whole point about digital signals is that they can be manipulated as we please providing that the underlying number codes are not altered in some irreversible way.

The specifications for the error-correcting system are:

Max. correctable burst error length is 4000 bits (2.5 mm length)

Maximum interpolable burst error length is 12 300 bits (7.7 mm length)

Other factors depend on the bit error rate, the ratio of the number of errors received to the number of bits total. The system aims to cope with bit error rates (BER) between 10^{-3} and 10^{-4}.

For BER = 10^{-3} interpolation rate is 1000 samples per minute, with undetected errors less than 1 in 750 hours

For BER = 10^{-4} interpolation rate is 1 sample in 10 hours with negligible undetected errors.

Last word

The proof of the efficacy of the whole system lies in the audio performance. The bare facts are impressive enough, with frequency range of 20 Hz to 20 kHz, (within 0.3 dB) and more than 90 dB dynamic range, signal-to-noise ratio and channel separation figures, with total harmonic distortion (including noise) less than 0.005%. A specification of that order with analogue recording equipment would be difficult, to say the least, and one of the problems of digital recording is to ensure that the analogue equipment which is used in the signal processing is good enough to match up to the digital portion.

Added to the audio performance, however, we have the convenience of being able to treat the digital signals as we treat any other digital signals. The inclusion of control and display data means that the number of items on a recording can be displayed, and we can select the order in which they are played, repeating items if we wish. Even more impressive (and very useful for music teachers!) is the ability to move from track to track, allowing a few notes to be repeated or skipped as required. The other consequence is that a CD can contain any data that can be put into digital form, so that it can carry sound, pictures and text. This is the basis of multimedia work, and what we have seen to date simply scratches the surface of what is possible.

NICAM

NICAM stereo sound is available on TV receivers in the UK, and is a method of broadcasting and receiving stereo sound for television. The system that has been used for FM is not suitable for TV because the additional bandwidth of the sound channel would overlap the TV bandwidth, so that a different method needs to be used. NICAM is an acronym for Near Instantaneous Companded Audio Method, and as the name suggests, it uses the digital methods of conveying two-channel signals by sending portions of the two channels alternately, as used for CDs.

All NICAM systems retain the normal sound carrier signal at 6 MHz above the vision carrier and carrying the mono signal, so that there is no compatibility problem for receivers which do not have the NICAM chips. At the transmitter, the two audio channels are sampled alternately at a rate of 32 kHz and with 14-bit digital output. This is then compressed to 10 bits, the digital equivalent of a type of Dolby process, and the usual parity bits are added. The signals are interleaved, as is done for CD (see Chapter 13), and assembled into 728 bit 'frames'. The digital signals are then phase-shift modulated on to a carrier which is at 6.552 MHz above the vision carrier. At the receiver these signals are separated from the frequency-modulated mono sound signals (by filtering), demodulated and separated into L and R channels using a storage chip so that both channels can be fed simultaneously once again. Finally the 10-bit signals are expanded to 14 bits again and converted to analogue form as the L and R audio outputs.

Camcorders

A camcorder is a combination of a TV camera, video recorder and synchronisation pulse generator, and the tiny modern camcorders are a triumph of miniaturisation, IC design, and design ingenuity. Though the picture quality does not match up to what can be attained using full-size professional equipment, the difference is nothing like as great as the price and size difference would suggest, and the poor quality pictures that are sometimes demonstrated are due more to the failings of the user rather than in the camcorder.

Camcorders are too small to use camera tubes, even the smallest types, so that they employ a form of light-sensitive chip that was originally developed for the TV cameras used in space exploration. The coding of video signals also uses methods that have been developed for space, and a valuable amount of bandwidth can be saved in this way, making it easier to record and replay on short lengths of tape (such as 8 mm or VHS-C) that are not practicable for the usual domestic systems. That said, the narrow 8 mm video tape system is used extensively in other countries – only the UK seems to be so attached to the old VHS standards.

Appendix A

Books for further reading

If you need more information on electronics, for design purposes, for constructional or for hobby interests, some of the books listed below may be of interest. In addition to these books, an immense amount of practical help is available from the catalogue of Maplin Electronics. The catalogues are available from Maplin stores, or by telephoning the main sales office whose number is (01702) 554161.

Some of the most useful reference books in electronics are either out of print or difficult to obtain. They have been included in the list below because they can often be found in libraries or in second-hand shops.

Components

Passive Components (Sinclair) Heinemann-Newnes 1990.
Manufacturer's Databooks by Texas, RCA, SGS-ATES, Motorola, National Semiconductor and Mullard contain detailed information on semiconductors, with many applications circuits.
Operational Amplifiers, 2nd edn (Jiri Dostal) Butterworth-Heinemann 1993.
Operational Amplifiers, 3rd edn (Clayton) Butterworth-Heinemann 1992.
Sensors and Transducers, 2nd edn (Sinclair) Butterworth-Heinemann 1992.
Switches (Sinclair) Butterworth-Heinemann 1988.

Computing

The Made Simple series (various authors), Butterworth-Heinemann.

Circuitry

Analog Circuit Design (Williams) Butterworth-Heinemann 1993.
The Art of Linear Electronics (Linsley Hood) Butterworth-Heinemann 1993.
Troubleshooting Analog Circuits (Bob Pease) Butterworth-Heinemann 1993.
Analogue Electronics (Hickman) Butterworth-Heinemann 1992.
Digital Logic Gates and Flip-flops (Sinclair) PC Publishing 1989.
Digital Logic Design, 3rd edn (Holdsworth) Butterworth-Heinemann 1993.
Digital Systems Reference Book (Holdsworth) Butterworth-Heinemann 1993.
Electronic Circuits Handbook, 2nd edn (Tooley) Butterworth-Heinemann 1992.
Maplin Electronic Circuits Handbook (Tooley) Butterworth-Heinemann 1990.
Newnes Electronic Circuits Pocket Book, Vol. I (Linear IC) (Marston) Butterworth-Heinemann 1991.
Newnes Electronic Circuits Pocket Book, Vol. II (Passive & Discrete circuits) (Marston) Butterworth-Heinemann 1993.

Formulae and tables

Reference Data for Radio Engineers (ITT) Howard Sams & Co. Inc. [Probably the most comprehensive data book ever issued.]
Newnes Radio and Electronics Engineer's Pocket Book, 16th edn, 1986, by

Keith Brindley. [Another excellent source of information, easier to get in the UK.]

Audio & Radio: Radio Designers' Handbook (Langford-Smith) Iliffe, 4th edn, 1967. [A wealth of data, though often on valve circuits. Despite the age of the book, now out of print, it is still the most useful source book for audio work.]

Electronic Engineers Reference Book, 6th edn (Mazda) Butterworth-Heinemann 1989. [A well-established standard reference book in the UK.]

Radio Amateurs Handbook (ARRL). [A mine of information of transmitting and receiving circuits. An excellent UK counterpart is available, but the US publication contains more varied circuits, because US amateurs are not so restricted in their operations.]

GE Transistor Manual (General Electric of USA). [Even the early editions are extremely useful.]

Microprocessors

Microprocessor Technology (Anderson) Butterworth-Heinemann 1994.
Practical Electronics Microprocessor Handbook, 1986, Ray Coles.
16-Bit Microprocessor Handbook, 1986, Trevor Raven.
The Art of Micro Design, 1984, by A. A. Berk.
Microprocessor Architectures and Systems (Heath) Butterworth-Heinemann 1993.
Microprocessor Data Book, 2nd edn (Money) Butterworth-Heinemann 1990.
Newnes 8086 Family (up to 80486) Pocket Book (Sinclair) Butterworth-Heinemann 1990.
Practical Microprocessor Interfacing (Money) Butterworth-Heinemann 1987.
32-Bit Microprocessors, 2nd edn (Mitchell) Butterworth-Heinemann 1993.

Electronics and computing topics

Linear Circuit Analysis and Drawing (Sinclair) Butterworth-Heinemann 1993 [A guide to the use of computers for linear circuit analysis and circuit drafting.]
Build Your Own PC (Sinclair) Butterworth-Heinemann 1994.
Servicing Personal Computers, 4th edn (Tooley) Butterworth-Heinemann 1993.
Newnes PC Troubleshooting Pocket Book (Tooley) Butterworth-Heinemann 1993.

Construction

Electronics Assembly Handbook (Brindley) Butterworth-Heinemann 1993.
Newnes Electronics Assembly Pocket Book (Brindley) 1992 Butterworth-Heinemann 1992.
Management of Electronics Assembly (Oakes) Butterworth-Heinemann 1992.
Soldering in Electronics Assembly (Judd) Butterworth-Heinemann 1992.

Servicing

Servicing Electronic Systems (Sinclair and Lewis) Avebury Technical 1991–6. [This is directed at the C&G 2224 Servicing course, and can also be used as a text of basic electronics theory and practice.]

Test equipment

Newnes Electronics Toolkit (Phillips) Butterworth-Heinemann 1993.
Modern Electronic Test Equipment, 2nd edn (Brindley) Butterworth-Heinemann 1990.
Oscilloscopes. How to Use Them, How They Work (2nd Edition), 1986, by Ian Hickman.

Appendix B

ASCII codes

These codes for alphabetical and numerical characters are almost universally used in computing and other digital applications. The core of code illustrated here covers the numbers 32 to 127, using a maximum of 7 binary bits. This permits the use of parity for error checking within a single byte (8 bits). Extended ASCII uses numbers 32 to 255, permitting a wide range of characters such as accented characters and other symbols, but these are not so strongly standardised. The best-known 8-bit set is the PC-8 set used on the IBM and clone computers.

The numbers have been given in both denary and binary forms. The table shows the UK ASCII set in which the £ sign is represented by 35. In the US set this number is used for the hash sign (#). The arrangement of numbers emphasises the relationships that are built into the table, such as the single-digit difference between a lower-case letter (such as **a**) and its upper-case equivalent (**A**).

Denary	Binary	Character	Denary	Binary	Character	Denary	Binary	Character
32	00100000	(space)	64	01000000	@	96	01100000	'
33	00100001	!	65	01000001	A	97	01100001	a
34	00100010	"	66	01000010	B	98	01100010	b
35	00100011	£	67	01000011	C	99	01100011	c
36	00100100	$	68	01000100	D	100	01100100	d
37	00100101	%	69	01000101	E	101	01100101	e
38	00100110	&	70	01000110	F	102	01100110	f
39	00100111	'	71	01000111	G	103	01100111	g
40	00101000	(	72	01001000	H	104	01101000	h
41	00101001	)	73	01001001	I	105	01101001	i
42	00101010	*	74	01001010	J	106	01101010	j
43	00101011	+	75	01001011	K	107	01101011	k
44	00101100	,	76	01001100	L	108	01101100	l
45	00101101	–	77	01001101	M	109	01101101	m
46	00101110	.	78	01001110	N	110	01101110	n
47	00101111	/	79	01001111	O	111	01101111	o
48	00110000	0	80	01010000	P	112	01110000	p
49	00110001	1	81	01010001	Q	113	01110001	q
50	00110010	2	82	01010010	R	114	01110010	r
51	00110011	3	83	01010011	S	115	01110011	s
52	00110100	4	84	01010100	T	116	01110100	t
53	00110101	5	85	01010101	U	117	01110101	u
54	00110110	6	86	01010110	V	118	01110110	v
55	00110111	7	87	01010111	W	119	01110111	w
56	00111000	8	88	01011000	X	120	01111000	x
57	00111001	9	89	01011001	Y	121	01111001	y
58	00111010	:	90	01011010	Z	122	01111010	z
59	00111011	;	91	01011011	[	123	01111011	{
60	00111100	<	92	01011100	\	124	01111100	|
61	00111101	=	93	01011101	]	125	01111101	}
62	00111110	>	94	01011110	^	126	01111110	~
63	00111111	?	95	01011111	_	127	01111111	(delete)

Index